T0202488

SpringerBriefs in History of Science and Technology

More information about this series at http://www.springer.com/series/10085

Michael Eckert

Establishing Quantum Physics in Munich

Emergence of Arnold Sommerfeld's Quantum School

Michael Eckert
Deutsches Museum
München, Germany

ISSN 2211-4564   ISSN 2211-4572 (electronic)
SpringerBriefs in History of Science and Technology
ISBN 978-3-030-62033-2   ISBN 978-3-030-62034-9 (eBook)
https://doi.org/10.1007/978-3-030-62034-9

This Springer imprint is published by the registered company Springer Nature Switzerland AG
The registered company address is: Gewerbestrasse 11, 6330 Cham, Switzerland

Foreword

The Quantum Network

The quantum revolution emerged from a series of crises of the classical mechanical worldview from the late nineteenth century to the early 1930s. This major transformation of physical knowledge was the subject of a research project funded by the Max Planck Society and pursued jointly by the Max Planck Institute for the History of Science and the Fritz Haber Institute (2006–2012).

The project focused on the long-term history of this process of knowledge restructuration, stressing the continuity of practices and structures. The development of quantum physics was a communal effort whose nature cannot be captured by a biographical approach focusing on just a few central figures: careful attention had to be paid to the broader community of researchers and to the network that enabled them to achieve what no single researcher could do alone. This challenge necessitated a large-scale collaborative research project. The work drew upon extensive archival records of correspondence, manuscripts, and notebooks. From its inception, the project undertook extensive efforts at establishing an international network of historians, philosophers, and physicists working on the history of quantum physics. The research results of the project have been published in various venues (see further readings below).

The emergence of quantum physics was triggered in part by conflicts between theoretical expectations and experimental results, but also importantly by the difficulty in integrating the major new physical theories of the nineteenth century—electrodynamics and thermodynamics—into the mechanical worldview. The paradigmatic example is blackbody radiation, which dealt with the thermodynamic properties of radiation enclosed in a container of mechanical atoms. A central role in this process of transformation was played by a small number of mental models, such as the quantum oscillator, the Bohr atom, or the light quantum. These mental models served a dual role of bundling a large and growing amount of empirical knowledge and posing the challenge to find a consistent theoretical description for this empirical knowledge, which would otherwise have been unrelated.

Central to this process of reorganization of knowledge was not only a large body of knowledge, ranging from the periodic system of elements to the thermal properties of matter and radiation, but also the persistence of certain theoretical structures and methods. Theoretical physicists were, therefore, confronted with critical decisions about which concepts and theoretical structures could be maintained in the emerging theory and could thus serve as a guide for the development of the theory.

One central research result of the project is the insight that the quantum revolution, even though it implied one of the most momentous transformations of classical physics since its beginnings, cannot be understood as a wholesale replacement of one conceptual system by another. Rather, quantum mechanics was the reinterpretation (Heisenberg's German term was *Umdeutung*) of the concepts of classical physics within a new framework, which left intact not only most of the established empirical knowledge, but also major parts of the formal structure of this conceptual framework.

In the quantum revolution some of these high-level and abstract structures survived, but it was frequently controversial how they were to be interpreted in their new context. Famous dissenters, such as Einstein and Schrödinger, while accepting the new theoretical structure, disagreed about its meaning and its connection to the traditional mechanical worldview. These disagreements have persisted up to this day, even though quantum mechanics by all counts is a highly successful predictive theory.

This process of transformation and reinterpretation involved a considerable number of physicists, chemists, and mathematicians from several countries. And unlike the case of the contemporaneous development of general relativity, a large number of these researchers made essential contributions to the development of the theory. The work of the eventual creators of quantum mechanics in its final form would not have been possible without the numerous previous achievements that their work was based on. What was the reason for this striking contrast in the trajectories of relativity and quantum theory?

It can hardly be argued that the resulting theories, whether mathematical or conceptual, were fundamentally different in their complexity. While the rise of quantum physics was based on a much larger amount of empirical data, it is not the case that the numbers of the involved scientists were required for experimental work. The major part of the empirical evidence that was relevant for the development already existed in the subdisciplines that prepared quantum theory. Central to the theoretical endeavor was rather a reorganization of the existing knowledge. Why then was a large number of scientists necessary for the creation of the new theory?

The hypothesis standing behind this series of books is that a multitude of perspectives and theoretical insights, dispersed over different fields of physics, mathematics, and chemistry, was necessary for the theoretical development. This process required the establishment of a network of researchers refocusing their work on quantum problems and actively engaging in a multi-centric debate about the emerging theory. In a relatively short time span following the first Solvay

Conference in 1911, a discernible subdiscipline of physics emerged: quantum theory. The small group of physicists that gathered in Brussels in 1911 shared an interest in the theoretical potential of the quantum of action, which had been proposed by Max Planck in 1900 and was quickly applied to a variety of physical phenomena. Yet, this concept was still controversial, and most participants saw it as little more than a specific theoretical tool within the confines of statistical mechanics and thermodynamics. By 1918, five years after Bohr had published his first papers on the quantum theory of the atom, it was widely accepted that quantum theory was not only the essential theoretical ingredient for atomic physics, but also that it offered hope for a new theoretical framework for the physics of matter and possibly also radiation.

This series of Springer Briefs, which opened with Arne Schirrmacher's *Establishing Quantum Physics in Göttingen: David Hilbert, Max Born, and Peter Debye in Context, 1900–1926* (2019), elucidates the institutional preconditions and constraints on the emergence of this network of quantum physicists, and on the role that this new research focus played in the transformation of existing institutions and the formation of new ones. This includes the effect of contemporary scientists' recognition of quantum problems, specifically the shifting of their research foci, reallocation of their resources, and reorganization of research structures and policies. While the emergence of new theoretical concepts and new experimental results have been extensively studied by historians of science, the interaction between institutional and intellectual developments still requires further study. The volumes in this series thus aim to bridge the gap between work on individual physicists' paths of discovery and research on the institutional history of science. Each volume deals with one of the major sites of exploration of quantum theory, where some of the most powerful and successful physicists and organizers of the work toward the elucidation of quantum problems lived: Göttingen, Berlin, Copenhagen, and, in this volume, Munich. The four books present very different institutional situations and research policies that formed the context for the shift of the research focus toward quantum physics.

The book you are now holding in your hands is a study on Arnold Sommerfeld's famous "nursery of theoretical physics" at the University of Munich. Here, Michael Eckert shows the importance of the personal and institutional networks for the emergence of quantum theory. Sommerfeld, originally a mathematician with little interest in theoretical physics, was a somewhat unlikely choice for a chair of theoretical physics when he was appointed in 1906. However, he quickly reoriented his research focus toward physics, including a keen interest in experimental research. By 1911, he had already become one of the earliest proponents of quantum theory, participating in the first Solvay conference and working on a quantum theory of X-rays. Through his elaboration of Bohr's quantum theory of the atom into a general theoretical framework and his standard work on atomic theory *Atombau und Spektrallinien*, which he frequently revised to reflect the fast development of atomic physics, he became one of the leading theorists of quantum physics. Possibly even more important for the development of quantum theory in the coming years was his exceptional talent as a charismatic teacher and prolific networker, which made Munich into a central hub in the fast-growing network of quantum physicists in the 1920s. It is no coincidence that the two most talented "child prodigies" of 1920s quantum physics, Wolfgang Pauli and Werner Heisenberg, were his students, nor that by the end of the decade "about a dozen of Sommerfeld's former disciples held chairs in theoretical physics," as Eckert writes in this volume.

In the case of Berlin, one is dealing in principle with a far more complex research landscape, which included the University, the Charlottenburg Polytechnic, Prussian Academy of Sciences, Imperial Physical Technical Institute, and the Kaiser Wilhelm Institute for Physics, explicitly founded for research in quantum physics. The book by Giuseppe Castagnetti and Hubert Goenner, *Establishing Quantum Physics in Berlin: Einstein and the Kaiser Wilhelm Institute for Physics (1917–1922)*, focuses on the Kaiser Wilhelm Institute, as it represented an early attempt to establish a dedicated research institute for the development of quantum physics within the wider academic landscape of Berlin. In their book, Castagnetti and Goenner show that the foundation of an institute under Einstein's directorship, promoted by an influential group of Berlin physicists, administrators, and industrialists, was a reaction to the increased differentiation within physics and to the increasingly important role of Planck's quantum as a germ of crystallization of a growing field of knowledge. The creation of the institute was a concrete reaction to this situation, which attempted to take advantage of Einstein's ability concerning conceptual integration for the organization of research. It turned out, however, that this attempt was essentially a failure. The integration of the knowledge from physics and chemistry that was required for the formulation of the later quantum mechanics could, it appears, neither be achieved by the intellectual work of a single outstanding individual such as Einstein nor by traditional types of intellectual cooperation. In particular, the failure of the Kaiser Wilhelm Institute to promote regular and lasting collaboration among scientists on a set of challenging research problems, across traditional disciplinary specializations, can be interpreted as a failure of the "academy model" of research policy and funding.

Alexei Kojevnikov's book on the history of Niels Bohr's research policies from 1913 to the advent of quantum mechanics in 1925 is based on extensive archival research in the Niels Bohr Archive in Copenhagen and in other archives. It not only shows that the political circumstances of the time provided a neutral country like Denmark with a special opportunity to play a new and more important role on the international scene. It also makes it clear that these circumstances, in particular after the First World War, weakened traditional scientific institutions and, at the same time, the traditional professional and intellectual loyalties fostered by them. As a consequence, the relatively small Bohr institute could create new, transnational bonds between scientists that were more flexible than the established ones. These new, less rigid connections also allowed for the integration of a variety of local scientific traditions, which were then brought to bear on specific problems of the emerging quantum theory. This development is exemplified by several case studies treated within the study. In the period covered by the study, Bohr's intellectual agenda evidently played an important role in the identification of research problems. But similar to the case of Sommerfeld in Munich, his organizational abilities as "a great politician, diplomat, fundraiser, and organizer of science" were at least as important in making Copenhagen a center of quantum physics. A central factor in this development was the emergence of a new professional stage for young researchers: the postdoctoral student. It was postdoctoral fellows on Rockefeller Foundation stipends that constituted a major and vital part of Bohr's institute, and that contributed in decisive ways to the development of quantum physics.

In the case of Göttingen, finally, studied by Arne Schirrmacher in the first book of this series, the institutional context is strikingly different from the other cases. After the Prussian takeover in 1866, Göttingen had become just another provincial university in the rather centralistic Prussian education system, and its relatively small physics faculty had to struggle with a shortage of resources and the disinterest of the Prussian ministry of education. Its mathematics department on the other hand was acknowledged as one of the, if not the, most important centers for mathematical research in Germany since the days of Carl Friedrich Gauss. By 1900, with Felix Klein and David Hilbert, it had even raised its profile and become internationally so important that even the Prussian bureaucracy could not ignore its demands. Hence, it is not surprising that the shift of the research focus to quantum physics in the years from 1914 to 1921, and the hiring of Peter Debye, Max Born, and James Franck in these years, happened not because of support from the existing physics faculty but mainly because of the interest of David Hilbert in the development of quantum physics. Schirrmacher shows how Hilbert reallocated resources toward quantum research by looking at his lectures, assistants, and use of funds. When Born was called to Göttingen in 1921, he already had established a successful research group on quantum physics in Frankfurt. Together with Franck, he quickly transformed Göttingen into a center of quantum physics, not only through their own work but also, again, through Born's network-building efforts, whereby he especially pursued the exchange with Sommerfeld in Munich and Bohr in Copenhagen. He was therefore able, for example, to have both Pauli and Heisenberg do their postdoctoral work in Göttingen, and also contribute greatly to the work of his institute during these years.

Let us mention, in closing, one striking common thread of the four books: in all cases, the success or failure of the program depended on the integration of the local institution into a network of quantum research. Munich, Copenhagen, and Göttingen could only be as successful as they were as research centers through their intensive exchanges. Berlin on the other hand, may have failed to develop as impressively exactly because of a more conservative and self-sufficient approach to research. This also may explain the relative lack of progress made in the well-established centers of research outside of central Europe, which were much less affected by the turmoil of the aftermath of World War I, since also collaboration and networking there could be seen as less of a necessity. This proposal would offer a striking alternative to Paul Forman's claim that it was the cultural climate of the Weimar Republic that pushed quantum physics to the fore. Taken together, the four volumes in this series present scientific institutions and their interconnections as the historical locus where the cultural climate of the times and the internal logic of the science meet. These networks of research thus provide an ideal starting point for transcending the old, artificial divide between external and internal histories of science.

Further reading in the context of this project includes the following publications: *Constructing Quantum Mechanics* by A. Duncan and M. Janssen (Oxford University Press, 2019); *Practicing the Correspondence Principle in the Old Quantum Theory* by M. Jähnert (Springer, 2019); "Special Issue: On the History of the Quantum, HQ4" by J. Navarro, A. Blum, C. Lehner, eds. *Studies in History and Philosophy of Modern Physics* (60:2017); *The Bumpy Road: Max Planck from Radiation Theory to the Quantum (1896–1906)* by M. Badino (Springer 2015); *Research and Pedagogy: A History of Quantum Physics through Its Textbooks* by M.Badino and J. Navarro, eds. (Edition Open Access, 2013); *Traditions and Transformations in the History of Quantum Physics: Third International Conference on the History of Quantum Physics, Berlin, June 28–July 2, 2010* by S. Katzir, C. Lehner, J. Renn, eds. (Edition Open Access, 2010); "On The History of The Quantum: The HQ2 Special Issue" by J. van Dongen et al (eds.), *Studies in History and Philosophy of Modern Physics* (40(4):2009).

Berlin, Germany   Christoph Lehner
June 2020   Jürgen Renn
                                                                                       Alexander Blum

Contents

1   **Introduction** 1
    References 3

2   **Boltzmann's Legacy** 5
    References 12

3   **Munich Beginnings** 15
    References 26

4   **X-rays and Quanta, 1911–1913** 27
    References 33

5   **Extending Bohr's Model, 1914–1919** 35
    References 50

6   **Synergy and Competition in the Quantum Network, 1919–1925** 53
    References 66

7   **Wave Mechanics—A Pet Subject of the Sommerfeld School,
    1926–1928** 67
    References 73

8   **Conclusion** 75
    References 76

Appendix 77

Abbreviations and Archives

AEA   Albert Einstein Archives, Hebrew University, Jerusalem
AETH   Archiv der Eidgenössischen Technischen Hochschule, Zurich
AIP   American Institute of Physics, College Park
AMPG   Archiv zur Geschichte der Max-Planck-Gesellschaft, Berlin
ARSA   Archive of the Royal Swedish Academy of Sciences, Stockholm
ASGS I-IV Arnold Sommerfeld. Gesammelte Schriften. I–IV (Sauter 1968)
ASWB I, II Arnold Sommerfeld. Wissenschaftlicher Briefwechsel I, II (Eckert and Märker 2000; Eckert and Märker 2004)
BANL   Biblioteca dell'Accademia Nazionale dei Lincei e Corsiniana, Rome
BayHStA   Bayerisches Hauptstaatsarchiv, München
CA   CalTech Archives, Pasadena
Caltech   California Institute of Technology
CUA   Catholic University Archives, Washington DC
DM   Deutsches Museum, Archiv, Munich
GOAR   Göttinger Archiv des Deutschen Zentrums für Luft- und Raumfahrt
GStA   Geheimes Staatsarchiv, Berlin
HA   Hochschularchiv, Rheinisch-Westfälische Technische Hochschule, Aachen
MBL   Museum Boerhaave, Leiden
NAS   Nobel Archive, Stockholm
NBA   Niels Bohr Archive, Copenhagen
NMAH   National Museum for American History, Smithsonian Institution, Washington DC
RANH   Rijksarchief in Noord-Holland, Haarlem
SBPK   Staatsbibliothek Berlin, Preußischer Kulturbesitz
SUB   Staats- und Universitätsbibliothek, Göttingen
UAL   University Archive, Universität Leipzig

UAM	University Archive, Ludwig-Maximilians-Universität, Munich
UAZ	University Archive, Zurich
ULM	University Library, Ludwig-Maximilians-Universität, Munich
WPWB I	Wolfgang Pauli. Wissenschaftlicher Briefwechsel I (von Meyenn 1979)

Chapter 1
Introduction

Abstract Arnold Sommerfeld's institute at Munich University formed the base of a quantum network that rapidly expanded, developing new nodes in Copenhagen and Göttingen. This book traces the history of this institute from the early beginnings around 1900 to the heyday of quantum mechanics in the 1920s. It is a story of haphazard twists and turns.

Keywords Arnold Sommerfeld · Munich · Quantum physics

"Hopefully, you will keep a nursery for physics babies, like Pauli and I had back in the day, open for a long time yet!" This congenial comment in a letter from Werner Heisenberg to Arnold Sommerfeld on the occasion of his sixtieth birthday shows what an impact Sommerfeld's institute had on the next generation of physicists.[1] The boyish beginnings of quantum mechanics, initiated by prodigies in the quantum schools of Sommerfeld in Munich, Niels Bohr in Copenhagen and Max Born in Göttingen, is a remarkable feature of the quantum revolution (Eckert 2001). Sommerfeld's school stands out in this regard.[2] "Atomic theory had grown up and developments in physics had taken place which made all of Sommerfeld's pupils go off, as the German saying is,'Wie warme Semmeln,' like hot rolls," one of Sommerfeld's pupils recalled. The 1920s was a time when practically every German university dedicated a chair or even an institute to theoretical physics.[3]

A quarter century earlier, in the late 1890s, the situation was quite different. Sommerfeld was then a professor of mathematics at the mining academy (*Bergakademie*) in Clausthal, with little interest in physics beyond its application to mathematics. His teacher, mathematician and entrepreneurial science organizer Felix Klein,

---

[1] Heisenberg to Sommerfeld, 6 February 1929, DM, HS 197728/A, 136."[…] hoffentlich halten Sie noch lange ein Erziehungsheim für Physikalische Babys wie für Pauli und mich seinerzeit!" All translations of German passages in the cited correspondence by Michael Eckert.

[2] Benz (1975), Eckert and Pricha (1984), von Meyenn (1993), Eckert (1993, 2013), Seth (2010).

[3] Interview with Ewald by Weiner, 17–24 May 1968. Oral Histories, Niels Bohr Library & Archives, American Institute of Physics (AIP), https://www.aip.org/history-programs/niels-bohr-library/oral-histories/4596-1, accessed 17 April 2020.

had just initiated a project to compile an *Enzyklopädie der mathematischen Wissenschaften* and assigned Sommerfeld the task of editing the volumes on physics—a chore that Sommerfeld tried to pitch to Wilhelm Wien. Wien had just made a name for himself with theoretical discoveries about black-body radiation (including Wien's displacement law and Wien's radiation law) that would later play a crucial role in the development of quantum theory. Sommerfeld was already busy with the projects of the entrepreneurial Klein. He suggested Wien as more competent for a task that involved theoretical physics. Wien, however, was well aware of the meager status of theoretical physics. "Theoretical physics does not find customers at present," he stated in his declination of the offer. As *professor extraordinarius* (associate professor) of physics at the Aachen Technical University, waiting for a *professor ordinarius* (full professor) position elsewhere, Wien complained "that physicists almost exclusively cultivate the pure experiment and are not interested in theory."[4] Faced with Wien's rejection, Sommerfeld assumed the chore himself. Ultimately, this task contributed significantly to Sommerfeld's shift from mathematics and sparked his unique career as one of Germany's first theoretical physicists. Wien, however, became an experimental physicist with reactionary tendencies—scientifically as much as politically. Once in 1922, for example, Max Planck criticized Wien's disdain of recent theoretical advances; these should not be regarded, as Planck said, "as superfluous luxury but as adamant consequences from new facts." Planck countered Wien's attitude, asserting that "the old views may not be retained entirely unchanged, whatsoever."[5] Sommerfeld's disciples would frequently experience the antagonism between Wien's and Sommerfeld's approaches toward physics. Heisenberg's near failure in his doctoral examination illustrates how far both tendencies diverged by the mid-1920s. Despite his expert knowledge in quantum theory fostered by Sommerfeld, Heisenberg had problems answering Wien's questions in the oral exam about basic subjects that were taught in elementary courses on experimental physics. He was allowed to pass only after some debate between Sommerfeld and Wien (Cassidy 2009, 117–119).

In retrospect, it seems ironic that Wien turned from a representative of contemporary theoretical physics on the eve of the quantum revolution circa 1900 into an experimental physicist with little sympathy for modern theories whereas Sommerfeld converted from a mathematician with little enthusiasm for theoretical physics into a missionary for this specialty—both directing institutes of experimental and theoretical physics under the umbrella of the same university in Munich. However, such twists merely illustrate that theoretical physics did not surface at once as a specialty

---

[4]Wien to Sommerfeld, 11 June 1898, DM, HS 1977–28/A,369. Also in ASWB I. "Die theoretische Physik liegt in Deutschland so gut wie vollständig brach. [...] Die Gründe dafür liegen erstens darin, daß die Physiker so gut wie ausschließlich das reine Experiment pflegen und für die Theorie kaum Interesse hegen[.] zweitens daran daß die meisten Mathematiker sich den ganz abstrakten Gebieten zugewandt haben [...] Die theoretische Physik findet gegenwärtig keine Abnehmer."

[5]Planck to Wien, 13 June 1922, Wien Papers, SBPK. "Neue Ideen sind aufgetaucht, nicht als überflüssiger Luxus, sondern als unerbittliche Folgerungen aus neuen Tatsachen, und die alten Anschauungen lassen sich nun einmal nicht ganz unverändert aufrechterhalten." Quoted in (Wolff 2008, 384).

in its own right. By the same token, the epitome of twentieth-century theoretical physics, quantum mechanics, was rooted in precarious institutional soil. Sommerfeld's and Wien's careers provide concrete examples for this growth process. Their trajectories crossed repeatedly during the three decades before the advent of quantum mechanics. The "sad story of Heisenberg's doctorate"[6] is only a late reminder that Sommerfeld's Munich quantum "nursery" was built on uncharted terrain.

This book traces the history of Sommerfeld's institute, beginning with the establishment of a chair for theoretical physics for Ludwig Boltzmann in 1890 up to the heyday of quantum mechanics in the 1920s. Boltzmann's legacy set high expectations that were not easily fulfilled at a time when theoretical physics, as its own discipline, was still in its infancy. Twists and turns like Wien's conversion from a quantum protagonist into a reactionary were not unusual at a time when disciplinary territories were still being carved out and new career paths were opening for ambitious theorists. When the "Bohr-Sommerfeld" atomic model emerged as a new item on the agenda of theoretical physics during the First World War, Sommerfeld dedicated considerable effort to its further exploitation. After the war, he turned quantum theory into a target of opportunity for a new generation of theorists. His institute formed the base of a quantum network that rapidly expanded, developing new nodes in Copenhagen and Göttingen. From an institutional perspective, the history of this quantum network reached a climax in 1928 when Sommerfeld's former prodigies, Heisenberg and Wolfgang Pauli, established new centers in Leipzig and Zurich, bringing quantum mechanics into the theoretical physics curricula on a large scale. By looking at the entire history of this development, we will encounter all kinds of contingencies that are easily overlooked if one considers only the outstanding achievements.

References

Benz U (1975) Arnold Sommerfeld: Lehrer und Forscher an der Schwelle zum Atomzeitalter, 1868–1951. Große Naturforscher, vol 38. Wissenschaftliche Verlagsgesellschaft, Stuttgart

Cassidy D (2009) Beyond Uncertainty. Heisenberg, Quantum Physics, and the Bomb. Bellevue Literary Press, New York

Eckert M (1993) Die Atomphysiker: Eine Geschichte der theoretischen Physik am Beispiel der Sommerfeldschule. Vieweg, Braunschweig

Eckert M (2001) The Emergence of Quantum Schools: Munich, Göttingen and Copenhagen as New Centers of Atomic Theory. Annalen der Physik 10:151–162

Eckert M (2013) Arnold Sommerfeld: Atomphysiker und Kulturbote 1868–1951. Eine Biografie. Göttingen: Wallstein. Abhandlungen und Berichte des Deutschen Museums, Neue Folge, Bd. 29. American Translation: Arnold Sommerfeld: Science, Life and Turbulent Times 1868–1951. Springer, New York

Eckert M, Pricha W (1984) Boltzmann, Sommerfeld und die Berufungen auf die Lehrstühle für theoretische Physik in Wien und München, 1890–1917. Mitteilungen der Österreichischen Gesellschaft für Geschichte der Naturwissenschaften 4:101–119

---

[6]This is the title of a chapter of an online exhibit on Heisenberg at http://www.aip.org/history/heisenberg/p06.htm, accessed 17 April 2020.

Seth S (2010) Crafting the Quantum. Arnold Sommerfeld and the Practice of Theory, 1890–1926. The MIT Press, Cambridge, MA

von Meyenn K (ed) (1993) Sommerfeld als Begründer einer Schule der Theoretischen Physik. In: Albrecht H (ed) Naturwissenschaft und Technik in der Geschichte. 25 Jahre Lehrstuhl für Geschichte der Naturwissenschaft und Technik am Historischen Institut der Universität Stuttgart. Verlag für Geschichte der Naturwissenschaft und der Technik, Stuttgart, pp 241–261

Wolff S (2008) Die Konstituierung eines Netzwerkes reaktionärer Physiker in der Weimarer Republik. Berichte für Wissenschaftsgeschichte 31:372–392

Chapter 2
Boltzmann's Legacy

Abstract The chair for theoretical physics at the University of Munich had been established in 1889 for Ludwig Boltzmann. It was abandoned when Boltzmann left Munich in 1894 until Wilhelm Conrad Röntgen, who in 1900 was called to the University of Munich as chair of experimental physics, insisted that theoretical physics be revived. In 1906, Sommerfeld was called to Munich as Boltzmann's successor.

Keywords Munich University · Ludwig Boltzmann · Wilhelm Wien · Wilhelm Conrad Röntgen · Hendrik Antoon Lorentz · Theoretical physics

When Wien declined Sommerfeld's offer to serve as physics editor for Klein's encyclopedia in 1898, he argued that theoretical physics was on sound footing only at the universities of Berlin and Göttingen. He could have added the University of Königsberg, where Sommerfeld studied and where, by the mid-nineteenth century, Franz Ernst Neumann and his school had founded an important tradition of theoretical physics (Olesko 1991). In the 1890s, Paul Volkmann, a disciple of Neumann, directed the theoretical physics institute in Königsberg. Neither Wien nor Sommerfeld, however, regarded Volkmann as a true representative of the field. Provoked by the news of Heinrich Hertz's untimely death in 1894 at the age of 36, Sommerfeld compared Hertz's physics with the physics he had learned at Königsberg. It appeared to Sommerfeld that his university physics professors were "wasters."[1] Wien considered only one other place besides Berlin and Göttingen worth mentioning: Munich. But in 1898, the approach to theoretical physics in Munich appeared more like a missed opportunity than a promising beginning. Wien even took it as an indication

---

[1]Sommerfeld to his mother, 5 January 1894. Private Sommerfeld Papers in the possession of Sommerfeld's heirs. I am grateful to Sommerfeld's heirs for the permission to quote from these papers. Henceforth, references to these papers are abbreviated as "Private Papers." "Hätte da nicht, wenn es gerade ein Physiker sein sollte, einer von den nichtsnutzenden Pape, Volkmann etc. darauf gehen können?".

of the decline of theoretical physics that "such an important chair like Munich has folded completely."[2]

The chair for theoretical physics at the university of Munich had been established only nine years earlier. In 1889, the dean of the philosophical faculty urged the Munich University Senate to elevate theoretical physics to a specialty in its own right, and argued that the need for it had already been felt "for a longer time." Furthermore, the dean strongly suggested that the time had now come to create a chair for theoretical physics "because at present there is an opportunity to find someone of the highest rank for this position." Boltzmann, at that time perhaps the best known theorist, was unhappy with his position as professor of experimental physics at the University of Graz in Austria, therefore it went without question "that he would accept a call to Munich as professor of theoretical physics with pleasure." A year earlier, Boltzmann was offered late Gustav Kirchhoff's chair of theoretical physics at the university in Berlin. Boltzmann first accepted but then declined the offer, apparently because of reservations about the Prussian way of life. An offer from Munich would not meet such resistance. The Munich faculty unanimously entreated the Senate not to miss this opportunity and to support their proposal "with urgency at the highest authority."[3]

The plan succeeded. In 1890, Boltzmann became a professor of theoretical physics at the University of Munich. His chair was combined with the position as curator for the Mathematical-Physical Collection of the Bavarian State under the authority of the Bavarian Academy of Science. Boltzmann perceived the latter obligation as a waste of time (Koch 1967). The instruments were later transferred to the German Museum (*Deutsches Museum*), but the dual responsibility as professor of the university and curator of the academy was a persistent tradition which continued to influence institutional affairs under Sommerfeld.

---

[2]Wien to Sommerfeld, 11 June 1898. DM, HS 1977–28/A, 369. Also in ASWB I. "Äußerlich zeigt sich das darin, daß reine theoretische Physik nur von zwei Lehrstühlen (Berlin und Göttingen) vorgetragen wird und ein so bedeutender Lehrstuhl wie München ganz eingegangen ist." For the rise of theoretical physics in Germany in the nineteenth century, see Jungnickel and McCormmach (1986).

[3]Von Bayer to the Senate of Munich University, 24 November 1889, UAM, Senats- und Dekanat-sakten, E-II-N, Boltzmann. "Wenn wir das hienach schon länger gefühlte Bedürfnis nach einer ordentlichen Professur der theoretischen Physik erst jetzt geltend machen, so geschieht dies, weil sich im gegenwärtigen Augenblick die sichere Gelegenheit bietet, eine Kraft ersten Ranges hierfür zu gewinnen. Es ist uns nämlich bekannt geworden, daß Dr. Ludwig Boltzmann, k.k. Regierungsrat, Professor der Experimentalphysik an der Universität Graz und Direktor des physikalischen Instituts daselbst (geb. zu Wien am 20. Februar 1844) den Wunsch hat diese Stellung mit einem Lehrstuhl der theoretischen Physik zu vertauschen, um sich mit voller Kraft seinem eigentlichen Arbeitsgebiet widmen zu können, und wir wissen, daß er einem Rufe nach München als Professor der theoretischen Physik gern folgen würde [...] Die gegenwärtige günstige Gelegenheit eine so hervorragende Lehrkraft für unsere Universität zu gewinnen, sollte nach unserer Ansicht nicht versäumt werden. Die Fakultät hat daher mit Einstimmigkeit beschlossen zu beantragen, daß Professor Dr. Ludwig Boltzmann in Graz als ordentlicher Professor der theoretischen Physik an unsere Universität berufen werde, und stellt an den k. Akademischen Senat die ergebenste Bitte, diesen Antrag an höchster Stelle dringend zu befürworten." See also Jungnickel and McCormmach (1986, 149).

It was not long until Boltzmann's performance in Munich yielded tangible results. In 1891 and 1893, his "Lectures on Maxwell's Theory of Electricity and Light" in Munich appeared in print (Boltzmann 1892b). Boltzmann also lectured "On the Methods of Theoretical Physics" during a public conference (Boltzmann 1892a). These lectures helped Munich become a center of theoretical physics as a specialty in its own right, comparable to Göttingen (with Woldemar Voigt) and Berlin (with Planck). However, in 1894, Boltzmann accepted a professorship in Vienna (Höflechner 1982, 52). The Munich faculty made only lukewarm attempts to replace Boltzmann, apparently in the hope that Boltzmann would soon return to Munich after he revealed that he was rather unhappy with the Viennese position. When Boltzmann indeed declared his wish to return to Munich two years later, however, the Ministry of Culture refused to call him back under the pretext of financial problems.[4] The budget for Boltzmann's chair was used "for other university purposes," as the dean of the faculty described the situation in 1900.[5] At this point, Saxony offered Boltzmann the newly created position of professor ordinarius of theoretical physics at the university of Leipzig. As had happened ten years earlier in Munich, the Leipzig faculty was motivated by the opportunity to attract his celebrity. And as before, Boltzmann's sojourn was short. He left Leipzig after only two years and returned to Vienna in 1902 (Jungnickel and McCormmach 1986, 174–180).

The origins of theoretical physics at Leipzig are beyond the scope of this study (Schlote 2004, 85–89). However, in addition to its similarities with Munich, there is another reason the Leipzig case deserves a closer look. "Perhaps you will have heard that I had hoped to receive you here as my colleague" Leipzig mathematician Carl Neumann wrote to Sommerfeld in May 1903, alluding to a quarrel in the Leipzig faculty about Boltzmann's successor. A tentative list of candidates included Sommerfeld, who was then regarded more as a mathematician than a physicist, but the experimental physicist Otto Wiener expected that a theoretical physicist should also be able to direct experimental work. The "salvation of physics," Neumann ridiculed of Wiener's attitude, required a theorist whose research gave rise to new experiments. Wiener's opinion prevailed and Sommerfeld's name was removed from the final list.[6]

It is no accident that Sommerfeld's name initially appeared on the Leipzig list. Neumann had performed groundbreaking work on potential theory, the subject of an article edited (and co-authored) by Sommerfeld in the *Enzyklopädie der mathematischen Wissenschaften* (Reiff and Sommerfeld 1904, Vol. 5, 3–62). Sommerfeld's correspondence with Neumann and others, about this and other articles, illustrates his growing attraction toward theoretical physics.[7] Some articles Sommerfeld edited addressed topical research of contemporary theoretical physics, such as those

---

[4]Ministry of Interior Affairs to Senate, 21 May 1896, Senats- und Dekanatsakten, UAM, OCI 22.

[5]Report to the Ministry of Interior Affairs, 24 February 1900, Senats- und Dekanatsakten, UAM, OCI 22. "[...] blieb die erledigte Stelle unbesetzt und der genannte Gehalt wurde bis auf eine Summe von wenigen Hundert Mark zu anderen Universitätszwecken verwendet [...]".

[6]Neumann to Sommerfeld, 22 May 1903. DM, HS 1977-28/A, 243. For more details about the Leipzig quarrel, see Schlote (2004, 87–88).

[7]See, for example, ASWB I, doc. 74–78.

authored by Hendrik Antoon Lorentz on electron theory and by Boltzmann on the kinetic theory of matter.[8] Wien also became a close colleague with whom Sommerfeld could consult for advice, despite his refusal to assume the editorial job himself.

Meanwhile, the situation in Munich was changing again. In 1900, Wilhelm Conrad Röntgen accepted the position as chair of experimental physics. The famous discoverer of X-rays, awarded the Nobel prize in 1901, complained to the Bavarian Ministry of Culture from the very beginning that his discipline would remain severely handicapped as long as the chair created for Boltzmann was deserted (Eckert and Pricha 1984). But initially, the complaints remained as futile as in preceding years, and Röntgen's past experience with the Bavarian Ministry of Culture did not give rise to optimism. "Bavaria's minister of culture is a dull bureaucrat," Röntgen had written to a friend when he was still at Würzburg University.[9] In Munich, he had little inducement to revise this impression. "Unfortunately, the circumstances in the Bavarian ministry of culture are not very favorable—at least with regard to the development of physics," he complained again in October 1904. Röntgen was then offered the directorship of the Imperial Physical Technical Institute (Physikalisch-Technische Reichsanstalt) in Berlin, the highest position a physicist could attain in Germany. In his negotiations with the Bavarian ministry, Röntgen left no doubt that he would follow the call to Berlin if Boltzmann's chair were not occupied by another first-rate theorist. "In doing so I have explicitly demanded nothing for *myself*," he revealed about his strategy.[10] The faculty considered it a "duty of honor" to support this effort.[11]

Faced with the risk of losing Röntgen, the Bavarian government released the required funds and authorized the faculty to fill Boltzmann's chair. Röntgen's preferred candidate was Lorentz, the leading authority on the theory of electrons, which Röntgen expected would explain the physical nature of X-rays and other phenomena in the fledgling field of mysterious rays. However, Lorentz declined the offer in Munich. At the request of the faculty, a commission was tasked with finding an appropriate candidate. In spring 1905, the commission's four members began an unprecedented effort to create a list of candidates. Twenty-one physicists were scrutinized for consideration. By June 1905, the list was narrowed to three names: Emil Cohn, Emil Wiechert and Sommerfeld. Both Cohn and Wiechert were authorities in the theory of electromagnetism (Darrigol 1995; Mulligan 2001). Sommerfeld had only recently begun to publish in this field; his name was brought to the attention of

---

[8]http://gdz.sub.uni-goettingen.de/en/dms/load/toc/?PPN=PPN360504019, accessed 17 April 2020.

[9]Röntgen to Zehnder, 4 November 1898 and 11 October 1904, quoted in Zehnder (1935, 69, 91–92). "Bayerns Kultusminister ist ein stumpfsinniger Bureaumensch."

[10]Röntgen to Zehnder, 11 October 1904, quoted in Zehnder (1935, 92). "Leider sind die Verhältnisse im Kultusministerium von Bayern keine—wenigstens für die Entwicklungen der Physik—sehr günstigen [...] Ich habe dabei ausdrücklich nichts für *mich* gefordert [...]".

[11]Faculty to Senate, 17 November 1904. UAM, OCI 31. For more details, see Eckert and Pricha (1984, 13–17), Jungnickel and McCormmach (1986, 275–277).

the commission by Boltzmann, Lorentz, and Wien, as was noted in the final report to the faculty. However, Sommerfeld's name was only in third place.[12]

The course of events which put Sommerfeld on the Munich list and finally led to his appointment as Boltzmann's successor again involves some ironic twists. When Lorentz declined the Munich offer, Röntgen asked him to recommend another candidate for this position: "We need no mathematician but a physicist who is equipped and familiar with the whole armory of mathematics and who knows exactly what physics needs" (Jungnickel and McCormmach 1986, 277). Sommerfeld certainly had an "armory of mathematics" at his disposal but no reputation of knowing "what physics needs"–as was apparent from the preceding Leipzig incident. He had just begun to publish on the theory of electrons–motivated by his work on Lorentz's encyclopedia article.[13] Lorentz must have praised Sommerfeld highly to persuade Röntgen that this candidate was not just another mathematician who ventured to apply his skills beyond the realm of his discipline.

Another irony concerns, once more, Wien's attitude. As the Munich commission noted, Wien brought Sommerfeld to their attention among other names. Less than three years earlier, Leipzig physicist Wiener quoted Wien as saying that Sommerfeld "would not be able to direct an institute and supervise students for work in physics" and therefore had to be taken off the Leipzig list.[14] However, starting in 1902 Sommerfeld's attitude was changing, and Wien had many opportunities to experience Sommerfeld's attraction toward physics. In February 1904, for example, Sommerfeld corresponded with Wien about the nature of X-rays.[15] Both Wien and Sommerfeld contributed to a *Festschrift* for Aachen physicist Adolf Wüllner. Sommerfeld wrote to Wien about the study he had undertaken for this purpose, "I perform almost as an experimental physicist," as it focused on a new method for measuring elastic constants of solids (Sommerfeld 1905). With regard to Wien's contribution on X-rays (Wien 1905), Sommerfeld suggested that they elaborate on certain features in which a comparison between theory and experiment would provide further information about X-rays.[16] "Isn't it a shame that ten years after Röntgen's discovery one still does not know more about the nature of X-rays," he wrote in another letter, praising Wien's contribution as one of the "most definitive and valuable con-

---

[12]Report to the faculty, 20 July 1905, UAM, E-II-N, Sommerfeld. "Von seiten sehr namhafter theoretischer Physiker wie Boltzmann, Lorentz und Wien wurden wir auf Sommerfeld aufmerksam gemacht, und es wurde die Erwartung ausgesprochen, daß er bei seiner hohen Begabung und seinem großen Fleiß noch viel leisten werde."

[13]Sommerfeld to Lorentz, 29 May 1904, RANH, Lorentz Papers. Also in ASWB I.

[14]Wiener, special report (*Sonderbericht*), 30 November 1902, UAL, PA 410. "[…] würde er nicht im Stande sein ein Institut zu leiten und Schüler zu physikalischen Arbeiten anzuleiten."

[15]Sommerfeld to Wien, 18 February 1904, DM, NL 56, 010. Also in ASWB I. Sommerfeld had analyzed the diffraction of X-ray "impulses" earlier as an extension of his habilitation concerning the mathematical theory of diffraction. See the references to these publications in ASWB I.

[16]Sommerfeld to Wien, 15 April 1905, DM, NL 56, 010. Also in ASWB I. The *Wüllner Festschrift* is available online at: http://archive.org/details/festschriftadol00wlgoog, accessed 17 April 2020. Sommerfeld was involved with the editing of the *Festschrift*.

tributions about X-rays."[17] In this exchange Sommerfeld displayed the attitude of a theorist who is more interested in the physics than in the mathematics of the problem. When the Munich commission scrutinized the candidates for Boltzmann's chair in the summer of 1905, Wien must have come to the conclusion that his earlier verdict about Sommerfeld was no longer true. Sommerfeld was apparently aware that it was Wien who brought his name to the attention of the Munich commission, because he asked Wien, "as the true cause of this event," for more information about the Munich chair. Röntgen had requested his "curriculum vita, list of publications etc. Something apparently is happening in Munich," he concluded.

> If you happen to know more about it, I would be very thankful for the information. [...] Would I have the opportunity to do experimental work at Röntgen's institute or should I be rather careful with such requests? Would I have an assistant in Munich or could I bring one with me?[18]

The fact that Sommerfeld was now asking for the possibility to do experiments was not the only ironic turn. Ferdinand Lindemann, the mathematician who supervised Sommerfeld's doctoral work in Königsberg, had been called to Munich in the meantime. Instead of supporting his former disciple, Lindemann expressed doubts about Sommerfeld's mathematical capabilities! Sommerfeld's recent papers on the theory of electrons, according to Lindemann, were mathematically flawed.[19] Lindemann's doubts also became part of Sommerfeld's evaluation: "Some of his works are probably not entirely flawless from a mathematical vantage point," the official report of the commission stated.[20] Once more, it was Wien who corrected a faulty verdict. Sommerfeld thanked him for his "efforts against the queer fellow Lindemann." Wien must have also informed Sommerfeld that he was placed behind Wiechert and Cohn because Sommerfeld responded with regard to Wiechert, his friend from Königsberg: "I hold Wiechert in such high esteem that it would not hurt me if he is preferred above me."[21] When Sommerfeld learned at the beginning of July 1906 that Wiechert would not accept an offer in Munich, he wrote Wien, as "the originator, supporter

---

[17]Sommerfeld to Wien, 13 May 1905, DM, NL 56, 010. Also in ASWB I. "Es ist eigentlich eine Schmach, daß man 10 Jahre nach der Röntgen'schen Entdeckung immer noch nicht weiß, was in den Röntgenstr. eigentl. los ist. Ihre Energiemeßung wird zu den sichersten und wertvollsten Kenntnißen über Röntgenstrahlen gehören."

[18]Sommerfeld to Wien, 4 July 1905, DM, NL 56, 010. Also in ASWB I "[...] als Sie wohl der Urgrund dieses Ereignißes sind. Röntgen telegraphirte Donnerstag an mich wegen Vita, Verzeichnis der Publikationen etc. Es scheint also in München etwas los zu sein. Sonst habe ich nichts gehört. Wenn Sie etwas Näheres wißen sollten, wäre ich Ihnen für Mitteilung sehr dankbar [...] Würde ich Gelegenheit zu gelegentl. experimenteller Arbeit im Röntgen'schen Institut haben oder muß ich mit solchen Anforderungen sehr vorsichtig sein? Würde ich einen Assistenten in München haben resp. mitbringen können?".

[19]Lindemann to Sommerfeld, 5 July 1905, DM, HS 1977-28/A, 203. In ASWB I. For the ensuing quarrel, see (Eckert 1997).

[20]Report to the faculty, 20 July 1905. UAM, E-II-N, Sommerfeld. "Einige seiner Arbeiten sind wohl vom mathematischen Standpunkt nicht ganz einwandfrei [...]."

[21]Sommerfeld to Wien, 5 November 1905, DM, NL 56, 010. Also in ASWB I. "Ich danke Ihnen herzlich für Ihre freundlichen Mitteilungen sowie für Ihre Bemühungen bei dem Querkopf Lindemann. Ich schätze Wiechert so hoch, daß es mir kein Schmerz ist, wenn er mir vorgezogen wird."

and protector of the idea to bring [him] to Munich," that he would probably "say yes" if he received the call.[22]

Three weeks later, the Bavarian ministry offered Sommerfeld the Munich chair. The conditions were largely the same as for Boltzmann, involving both the title professor ordinarius of theoretical physics and the position as curator of the mathematical-physical state collection. Additionally, they offered some interesting prospects for the future, namely, an institute for theoretical physics. The annual salary would be 6,900 Marks. His responsibility for the mathematical-physical state collection involved additional personnel: "an assistant with a salary of about 1,200 Marks in the beginning, and a mechanic, who would also be an attendant, with an initial salary of 1,500 Marks and a supplementary 255 Marks, the annual budget of the collection is 1,800 Marks."[23]

Sommerfeld negotiated for minor improvements concerning his budget, work space and salary–but left no doubt that he would accept the Munich offer.[24] "I cordially have rejoiced about the news that you receive the call to Munich and accept it," congratulated David Hilbert, who knew Sommerfeld during his study in Königsberg and was aware, like a few others, about the twists and turns of his career between mathematics and physics. "There you will be in touch with physics, the mother of all sciences, and feel happy in its womb."[25] Prior to the official notice, Sommerfeld also informed the Prussian ministry about his decision: "I intend to follow the call

---

[22] Sommerfeld to Wien, 5 July 1906, DM, NL 56, 010. Also in ASWB I. "Da ich Sie als den Urheber, Förderer und Beschützer der Idee vermute, mich nach München zu verpflanzen, so eile ich Ihnen zu melden, dass Wiechert den Münchener Ruf abgelehnt hat, wie mir seine Mutter nach früherer Verabredung telegraphirt. Näheres weiss ich noch nicht. Wenn also mein Freund Lindemann nicht eine energische Gegenaktion betreibt, so ist es wohl nach Lage der Dinge nicht unwahrscheinlich, dass ich gefragt werde und Ja sage."

[23] Von Wehner to Sommerfeld, 23 July 1906, DM, NL 89, 019, Mappe 5,2. "Mit der Professur ist auch die Stelle eines Konservators (:Vorstandes:) der mathematisch-physikalischen Sammlung des Staates verbunden. Der Gehalt von 6900 M gilt für beide Stellen. [...] In dem Neubauprojekte zur Erweiterung der Unversität, mit dessen Ausführung demnächst begonnen werden wird, sind eigene Räume [...] für ein Institut für theoretische Physik vorgesehen. [...] Das Personal bei der mathematisch-physikalischen Sammlung besteht aus einem Assistenten mit rund 1200 M Anfangsgehalt und einem Mechaniker, zugleich Diener mit einem Anfangsgehalte von 1500 M und 255 M Zulage, der Realetat der Sammlung beträgt 1800 M jährlich [...] Die Ernennung auf die Professur ist mit Wirkung vom 1. Oktober l. Js. an in Aussicht genommen."

[24] Sommerfeld to the Ministry of Culture, 12 August 1906, BayHStA, MK 11317; Assignment Certificate, 8 September 1906, DM, NL 89, 016, Mappe 1,7; Sommerfeld to Wien, 12 September 1906, DM, NL 56, 010. "In München hat man mir bereitwilligst die Einrichtung der Akademie-Räume mit Strom etc. zu Beobachtungszwecken bewilligt; auch noch 500 M zum Gehalt zugelegt."

[25] Hilbert to Sommerfeld, 29 July 1906, DM, HS 1977-28/A, 141. "Die Nachricht, dass Sie den Ruf nach München erhalten haben und annehmen, hat mich aufs herzlichste gefreut. [...] Sie kehren dort ein bei der Physik, der Mutter aller Wissenschaften, in deren Schoss Sie sicher sich glücklich fühlen werden."

because I consider theoretical physics as my true field of work." With regard to the salary, he assumed that he would not be "considerably better off" in Munich than in Aachen.[26] One of his Aachen engineering students recalled that Sommerfeld had told him once: "I am not really a technical professor, I am a physicist."[27]

References

Boltzmann L (1892a) Über die Methoden der theoretischen Physik. In: von Dyck W (ed) Katalog mathematischer und mathematisch-physikalischer Modelle, Apparate und Instrumente. Wolf und Sohn, Munich, pp 89–98. With addendum 1893

Boltzmann L (1892b) Vorlesungen über Maxwells Theorie der Elektricität und des Lichtes: I. und II. Teil. Graz: Akademische Druck- und Verlagsanstalt. Originally at Leipzig: Barth, 1891 and 1893. Reprinted, commented and annotated by Walter Kaiser in: Ludwig Boltzmann Gesamtausgabe (ed. Roman U. Sexl), Vol. 2

Darrigol O (1995) Emil Cohn's Electrodynamics of Moving Bodies. Am J Phys 63:908–915

Eckert M (1997) Mathematik auf Abwegen: Ferdinand Lindemann und die Elektronentheorie. Centaurus 39:121–140

Eckert M, Pricha W et al (1984) Geheimrat Sommerfeld—Theoretischer Physiker. Deutsches Museum, Munich

Höflechner W (1982) Ludwig Boltzmann: Sein akademischer Werdegang in Österreich. Dargestellt nach archivalischen Materialien. Mitteilungen der Österreichischen Gesellschaft für Geschichte der Naturwissenschaften 2:43–62

Jungnickel C, McCormmach R (1986) Intellectual Mastery of Nature: Theoretical Physics from Ohm to Einstein, vol 2. The University of Chicago Press, Chicago

Koch E-E (1967) Das Konservatorenamt und die Mathematisch-Physikalische Sammlung der Bayerischen Akademie der Wissenschaften. Unpublished working report, Institut für Geschichte der Naturwissenschaften der Universität München

Mulligan JF (2001) Emil Wiechert (1861–1928): Esteemed Seismologist, Forgotten Physicist. Am J Phys 69:277–287

Olesko K (1991) Physics as a Calling: Discipline and Practice in the Königsberg Seminar for Physics. Cornell University Press, Ithaca

Reiff R, Sommerfeld A (1904) Elektrizität und Optik: Standpunkt der Fernwirkung. Die Elementargesetze. In: Enzyklopädie der mathematischen Wissenschaften, vol 5, Part 2, Chap 12, pp 3–62. http://resolver.sub.uni-goettingen.de/purl?PPN360709672 (visited on 04/17/2020). Paper completed in December 1902; published in 1904

Schlote K-H (2004) Zu den Wechselbeziehungen zwischen Mathematik und Physik an der Universität Leipzig in der Zeit von 1830 bis 1904/05. Abhandlungen der Sächsischen Akademie der Wissenschaften zu Leipzig, Mathematisch-naturwissenschaftliche Klasse 63:1

Sommerfeld A (1905) Lissajous-Figuren und Resonanzwirkungen bei schwingenden Schraubenfedern; ihre Verwertung und Bestimmung des Poissonschen Verhätnisses. In: Festschrift Adolph Wüllner gewidmet zum siebzigsten Geburtstage 13. Juni 1905 von der Königl. Technischen Hochschule zu Aachen, ihren früheren und jetzigen Mitgliedern. Teubner, Leipzig, pp. 162–193. http://www.archive.org/details/festschriftadol01wlgoog (visited on 11/29/2020)

---

[26]Sommerfeld to Naumann, 29 July 1906. GStA, I. HA, Rep. 121 D II, Sekt. 6, Nr. 10. "Ich beabsichtige dem Rufe zu folgen, da ich in der theoretischen Physik mein eigentliches Arbeitsgebiet sehe und die Münchener Tätigkeit mich besonders anzieht. Pekuniär werde ich, wie es scheint, nicht wesentlich besser wie in Aachen gestellt sein."

[27]Rummel to Sommerfeld, 3 August 1906, DM, NL 89, 012. Also in ASWB I. "Sie haben mir ja sclbst einmal gesagt: 'Ich bin ja eigentlich kein technischer Professor, ich bin Physiker!'".

Wien W (1905) Über die Energie der Kathodenstrahlen im Verhältnis zur Energie der Röntgen- und Sekundärstrahlen. In: Festschrift Adolph Wüllner gewidmet zum siebzigsten Geburtstage 13. Juni 1905 von der Königl. Technischen Hochschule zu Aachen, ihren früheren und jetzigen Mitgliedern. Leipzig: Teubner, pp 1–14. http://www.archive.org/details/festschriftadol01wlgoog (visited on 11/29/2020)

Zehnder L (ed) (1935) W. C. Röntgen: Briefe an L. Zehnder. Rascher, Zurich

Chapter 3
Munich Beginnings

Abstract Together with his assistant, Peter Debye, Arnold Sommerfeld was eager to create in Munich, with seminars and colloquia, a "nursery of theoretical physics." Within a few years, he had developed an esteemed reputation for his novel, ambitious, and unconventional pedagogical practices. These extended to weekend trips in the Alps nearby, and, after colloquia, to bowling sessions in the basement of a local Munich beer hall. The scope of his research was wide and gave rise to a broad range of themes for doctoral dissertations. Apart from Debye, the early "Sommerfeld school" before World War I included Ludwig Hopf, Paul Peter Ewald, and Wilhelm Lenz.

Keywords Sommerfeld school · Peter Debye · Ludwig Hopf · Paul Peter Ewald · Wilhelm Lenz

In Munich, Sommerfeld felt that his dearest dreams had finally come true. As early as 1892, he had marveled about a holiday in Munich at the end of his studies: "I have a heavenly impression of Munich. That I didn't study here!!"[1] He loved the city, its museums, the Bavarian way of life, the Alps within close reach; Munich was so dear to him that he and his fiancée decided months before the wedding to honeymoon in Munich.[2]

While Sommerfeld had immediately fallen in love with the location of his new professional activity, his conversion to theoretical physics came rather gradually. In this regard, his first love was mathematics. When he agreed (under precarious conditions) to be Klein's assistant in 1894, he turned down an offer from Voigt to become an assistant at the Göttingen institute of theoretical physics because he did not want to dedicate his career to something "which I do not wholeheartedly consider as my

---

[1] Sommerfeld to his parents, 25 August 1892, Private Papers. "Ich habe einen himmlischen Eindruck von München bekommen. Dass ich hier nicht studiert habe!!".

[2] Sommerfeld to his parents, 26 October 1897, Private Papers. "Hochzeitsreise geht nach München".

M. Eckert, *Establishing Quantum Physics in Munich*,
SpringerBriefs in History of Science and Technology,
https://doi.org/10.1007/978-3-030-62034-9_3

mission."[3] He perceived the subject of his habilitation, the theory of diffraction, as a mathematical subject and chided Gustav Kirchhoff for not properly dealing with this topic as a physicist.[4] In 1899, Sommerfeld still hoped to find a permanent professional home in mathematics, preferably (alongside Klein and Hilbert) at the Mecca of this discipline: Göttingen.[5] As a representative of technical mechanics in Aachen, he took to engineering concerns with such enthusiasm and success that he received invitations to professorships at other technical universities.[6] When he learned that he was under consideration for Boltzmann's chair in Munich, he expressed his joy about this honor, yet remarked that he would leave Aachen with a heavy heart.[7] Theoretical physics had only recently become his true calling as a result of his interaction with Lorentz, Wien and others concerning their articles for the *Enzyklopädie der mathematischen Wissenschaften.* By then, his career had been contingent on the circumstances and opportunities in the uncharted disciplinary territory between mathematics and physics.

However, as soon as Sommerfeld arrived in Munich, he was eager to establish himself as a theoretical physicist in his own right—conscious about the tradition founded by Boltzmann 16 years earlier. Sommerfeld brought with him Peter Debye, who had come to Aachen as an engineering student and became Sommerfeld's assistant in 1904.[8] Despite their official responsibilities for engineering concerns, both Sommerfeld and Debye shared a vital interest in electron theory and other topics of theoretical physics. Debye, a native Dutchman, helped Sommerfeld present a recent paper on electron theory to the Amsterdam Academy of Science. The call to Munich offered both men an opportunity to pursue theoretical physics henceforth on a larger scale. "From the very beginning, I have spared no effort and attempted to create a nursery of theoretical physics in Munich by establishing seminars and colloquia."[9]

---

[3] Sommerfeld to his parents, 27 June 1894, Private Papers. "Ich habe bei Voigt gestern abgelehnt (hoffentlich hat er es nicht übel genommen!) (1) komme ich wieder in eine schiefe Stellung mich mit Dingen zu beschäftigen, die ich doch nicht ganz für meine Aufgabe ansehe. (2) kann die Klein'sche Assistentenstelle zu Okt. hier noch frei werden, jedenfalls aber Oktober über's Jahr."

[4] Sommerfeld to his mother, 3 October 1894, Private Papers. "Zudem macht mir der Herr Kirchhoff Sorge. Ich habe die gegründete Ansicht, dass das Alles Humbug u. Redensarten sind, was dieser mathematisch gründlichste unter den Physiker[n] in der Optik gemacht hat."

[5] He hoped, for example, to succeed Arthur Schönflies, at the time professor extraordinarius of applied mathematics in Göttingen, who was called in 1899 to Königsberg. Schönflies to Sommerfeld, 20 September 1899, DM, HS 1977–28/A,311.

[6] Mining Academy (*Bergakademie*) Berlin to Sommerfeld, 29 October 1904, DM, NL 89, 019, Mappe 5,3; J. Cardinaal (TH Delft) to Sommerfeld, 4 July 1906, DM, HS 1977–28/A,48.

[7] Sommerfeld to Wien, 4 July 1905, DM, NL 56, 010. Also in ASWB I. "Übrigens laßen Sie alle diese Fragen unbeantwortet, wenn es gegen die Discretion ist. Es würde mir schwerer werden von Aachen fortzugehn als man glauben sollte. Meine Tätigkeit und Stellung hier ist sehr angenehm und die Aachener Lebensbedingungen äußerst erfreulich."

[8] Sommerfeld to Borchers, 12 December 1904, HA, 844. "Seit Ablegen seines Vorexamens hat sich Herr Debye in äusserst erfolgreicher Weise privatim in die höheren Teile der Mechanik und theoretischen Physik hineingearbeitet."

[9] Sommerfeld to the Academy of Science in Vienna, 1919, AMPG, III. Abt., Rep. 19 (Debye). Later (1950) extended and published as "Autobiographische Skizze" in ASGS IV, 673–682, here p. 677.

Sommerfeld's ambition, as expressed in this memory 13 years later, seems to have outlasted or displaced the recollection of obstacles he and Debye had to cope with immediately after their arrival in Munich in the fall of 1906. Being ranked only third on the Munich faculty search list, Sommerfeld first had to meet Röntgen's high expectations for Boltzmann's successor and gain his trust—a task that was likely not facilitated by Röntgen's unapproachable reputation (Fölsing 1995). Another concern dealt with the equipment of his new institute, which was under construction in a new wing of the university. He would have to ensure that, even as a theorist, ample space would be provided for experimental physics, particularly in light of the Leipzig incident. "Is it advisable to use stone pillars?" he asked Wien, who had meanwhile become his trusted advisor, "Maybe you have some experience about stone pillars from your stay at the Reichsanstalt."[10] The construction of his institute was accomplished only three years later. For the time being, Sommerfeld and his assistant were stationed some distance from the university in a building that belonged to the Bavarian Academy of Science and housed the State Collection.

Lindemann's claim that Sommerfeld's electron theory was flawed was a more urgent concern. Röntgen, for his part, must have overcome any lingering doubts rather quickly. "I am very happy with Röntgen," Sommerfeld wrote in another letter to Wien. "He communicates with me scientifically and officially in a most friendly manner."[11] Röntgen expressed the same feeling in his private correspondence. "I believe that I have found a good colleague and collaborator in Sommerfeld," he confided to a friend. "I can speak about physical subjects again in an inspiring manner, and the students are interested in his lecture on Maxwell's theory and the theory of electrons." Sommerfeld did not always agree with him, as Röntgen added, but he considered this to be a positive trait.[12] Lindemann, however, insisted that Sommerfeld had fooled the physicists with his electron theory. The controversy was pursued privately through correspondence and publicly in the form of communications with the Bavarian Academy. But it turned out that Lindemann's reproach of Sommer-

---

"Es war für mich und meinen damaligen Assistenten Debye selbstverständlich, daß dieser Ruf uns beiden galt, d.h. daß Debye mich nach München begleitete. In München kam ich zum ersten Male dazu, Vorlesungen über die verschiedenen Gebiete der theoretischen Physik und Spezialvorlesungen über die im Fluß befindlichen Fragen zu halten. Ich habe von Anfang an dahin gestrebt und habe es mich keine Mühe verdrießen lassen, in München durch Seminar- und Colloquiumbetrieb eine Pflanzstätte der theoretischen Physik zu gründen."

[10]Sommerfeld to Wien, 7 January 1907, DM, NL 56, 010. "Bei dem Neubau der Universität sind Räume für ein theoretisch-physikalisches Institut vorgesehen [...]. Hat es nun einen Zweck, Steinpfeiler von unten her in's Hochparterre zu führen? [...] Vielleicht haben Sie auch aus der Reichsanstalt her über Steinpfeiler Erfahrungen?".

[11]Sommerfeld to Wien, 23 November 1906, DM, NL 56, 010. "Zunächst bin ich über Röntgen sehr glücklich. Er kommt mir wissenschaftlich und amtlich äußerst freundlich entgegen [...]".

[12]Röntgen to Zehnder, 27 December 1906, quoted in Zehnder (1935, 112). "An Sommerfeld glaube ich einen guten Kollegen und Mitarbeiter gefunden zu haben. Ich kann auch wieder in anregender Weise über physikalische Dinge reden, und die Zuhörer interessieren sich sehr für seinen Vortrag, über die Maxwellsche und über die Elektronentheorie. Wir sind nicht immer über die da vorkommenden Fragen einig; aber das schadet ja nichts: im Gegenteil, das dürfte die Sache und auch das Verständnis fördern können."

feld was based on Lindemann's own mathematical errors. The once highly respected mathematician became the target of ridicule and, even worse, pity: "So the result seems to be that Lindemann, due to a lack of an appropriate physical understanding, trusts in mere calculations and is misled himself because of frequent calculational errors," Klein summarized of the controversy's the outcome. "In view of the great talent which Lindemann undoubtedly originally showed, a tragic end."[13] The trouble with Lindemann could be considered mere interference from someone who did not belong to the physics community—and who was defeated finally by scientific arguments. Lindemann remained resentful of the physicists until the end of his career. When he resigned many years later, he explained to the minister, "I do not not want to fill the evening of my life with bitterness as a result of being officially chained to the gentlemen Wien and Sommerfeld."[14]

But Sommerfeld was also met with distrust from physicists. During the interim between Boltzmann's departure in 1894 and Sommerfeld's arrival in 1906, theoretical physics was taught by Leo Graetz whose Munich career had begun in the 1880s as a *Privatdozent* with Planck. By 1906, he had attained the rank of professor extraordinarius. Graetz had more than two decades of experience as a lecturer on theoretical physics, but Röntgen did not seriously regard him as a candidate for Boltzmann's chair (Eckert and Pricha 1984, 16; Jungnickel and McCormmach 1986, 276). At the age of fifty, Graetz was in an awkward situation when Sommerfeld, twelve years younger, was assigned the task of elevating the status of theoretical physics in Munich. Even Röntgen admitted that Graetz was now "somewhat sidelined" and proposed to the ministry that Graetz be awarded "the title and rank, but not the rights of an ordinary professor of physics."[15] Röntgen's proposal materialized, but it did not really sweeten Graetz's displacement. Without the means of an ordinary professorial chair, Graetz depended on the mercy and goodwill of Röntgen. He continued to lecture and successfully authored a textbook, but after Sommerfeld's arrival he must have felt like an outdated remainder of another era. Twenty years later, in a review of a century of physics Graetz recalled, "Among the theoretical physicists there are some who are skilled calculators and who know how to apply self-imposed theorems to special problems." Without naming Sommerfeld he left no doubt that such "mathematicians disguised as physicists" were not to his

---

[13]Klein to Sommerfeld, 20 November 1907, DM, HS 1977–28/A, 170. "Lindemann geht mir außerordentlich nahe. Das Resultat wäre also, daß Lindemann mangels geeigneter physikalischer Anschauung sich auf das bloße Rechnen verläßt und da in Folge gehäufter Rechenfehler in die Irre geht! Bei der zweifellos von Hause aus außerordentlich hohen Begabung von Lindemann ein tragisches Ende." For more detail, see Eckert (1997, 132).

[14]Lindemann to the Minister of Culture, 30 June 1923, BayHStA, MK 17841. "[…] denn ich habe nicht die Absicht, mir meinen Lebensabend dadurch verbittern zu lassen, dass ich mit den Herren Wien und Sommerfeld dienstlich zusammengekettet bin."

[15]Röntgen to Zehnder, 27 December 1906, quoted in Zehnder (1935, 112). "Graetz ist nun was seinen Lehrauftrag betrifft, 'Teilnahme an der Leitung des Praktikums und Verpflichtung, regelmäßige Vorlesungen zur theoretischen Physik zu halten', etwas kaltgestellt, und das ist, wie man auch sonst über ihn denken mag, für ihn in seinem Alter eine nicht angenehme Erfahrung. Um ihm die Veränderung seiner Stellung etwas zu versüßen, wurde dem Ministerium vorgeschlagen, ihm den Titel und den Rang, aber nicht die Rechte eines ordentlichen Professors für Physik zu verleihen."

liking (Graetz 1926).[16] Among Sommerfeld's circle, "Grätz" became subject of a wordplay—alluding to the German word *Krätze* for scum. "If he is your only Grätz you should congratulate yourself," Sommerfeld once consoled Debye, his former assistant, about an unpleasant colleague.[17]

Arthur Korn was another lecturer who presented courses on theoretical physics as a *Privatdozent* from 1895 and as professor extraordinarius beginning in 1903. Like Graetz, he regarded Sommerfeld's appointment as a slight to his own career. Unlike Graetz, however, Korn did not resentfully withdraw to a secondary position. When the faculty denied him the chair of applied mathematics as compensation, Korn provoked his dismissal and started a new career in Berlin. He also expressed his anger publicly in the *Berliner Tageblatt*. "In 1906, someone from outside was named the chair of theoretical physics," he explained of the origin of the feud. "Their ignoring me was, of course, a snub for me [...] The reason why the faculty left me in the lurch is without a doubt Professor Röntgen. He simply tyrannized the faculty in this matter" (Litten 1993, 46).[18]

Wilhelm Donle was a third lecturer on theoretical physics in Munich. He had become a Privatdozent at Munich University in 1888. Although he later served as a teacher in military high schools, he continued to present courses at the university. During Sommerfeld's early Munich years, Donle was a professor at the Royal Artillery and Engineering School (*Kgl. Artillerie- und Ingenieursschule*) with the rank and salary of a professor ordinarius.[19] However, as a lecturer with a rather loose affiliation with the university's physics institute, Donle and his lectures seem not to have been regarded as interfering with Sommerfeld's fledgling effort to establish theoretical physics as a specialty of its own right.

Despite the mutual resentments between the Munich professors, a student with an interest in theoretical physics could choose from an impressive array of lectures even before Sommerfeld's arrival in 1906 (Table A.1). Sommerfeld did not have to start "from scratch." After Boltzmann vacated his chair in 1894, Munich continued to offer lectures in theoretical physics to which Sommerfeld added his own. Sommerfeld's lecture model evolved over time to become the now canonical method of teaching theoretical physics. Lasting six semesters and four hours a week, the core lecture course consisted of classes on mechanics, mechanics of continuous media, electrodynamics, optics, thermodynamics, and partial differential equations. In addi-

---

[16]"Unter den theoretischen Physikern gibt es eine Klasse, die geschickte Rechner sind, und welche die selbstgestellten Sätze auf spezielle Einzelprobleme anzuwenden verstehen. [...] es sind Mathematiker in physikalischer Verkleidung; sie werden gewöhnlich von den Mathematikern als gute Physiker und von den Physikern als gute Mathematiker angesehen."

[17]Sommerfeld to Debye, 6 August 1920, AMPG, III. Abt., Rep. 19 (Debye). Also in ASWB 2. "Wenn er Dein einziger Grätz ist, so kannst Du Dir Glück wünschen."

[18]"Im Jahre 1906 wurde die ordentliche Professur für theoretische Physik [...] durch eine Berufung von außerhalb neu besetzt. Dadurch, daß man mich überging, wurde ich natürlich vor den Kopf gestoßen [...] Die Ursache, warum die Fakultät mich im Stich gelassen hat, ist zweifellos Professor Röntgen. Er hat in der ganzen Angelegenheit einfach die Fakultät tyrannisiert."

[19]Unpublished biographical material and further references on Donle are available in German at http://litten.de/fulltext/donle.htm, accessed 17 April 2020.

tion, students chose an elective on an advanced topic, which lasted two hours a week (Table A.4).

Sommerfeld soon developed an esteemed reputation for his novel, ambitious, and unconventional pedagogical practice. Just as he had focused on the needs of engineering students at the Technical University in Aachen six years prior, he now eagerly oriented his approach to physics. Many years later, Abram Ioffe, a regular collaborator of Röntgen's at that time, recalled how Sommerfeld approached him about familiarizing himself with the physics routine in Röntgen's institute, and for this reason he requested permission to visit Ioffe's laboratory two hours daily. Ioffe instead suggested that Sommerfeld join their group at the Hofgarten cafe where they informally met with chemists and crystallographers for "shoptalk" about their mutual work. "With his characteristic assiduousness, Sommerfeld appeared each day for about an hour," Ioffe recalled of the newcomer at their coffee table. "Sometimes it was not easy for Sommerfeld to follow our arguments, but his assistant Debye soon surpassed all of us" (Ioffe 1967, 39).[20] Ioffe's recollection fits with evidence from contemporary correspondence. In 1907, Sommerfeld remarked in a letter to a mathematician that he was now "concerned with the more real and special problems of physics rather than the general mathematical problems with which you are dealing."[21] He also came into closer contact with experimental physicists. In May 1907, for example, he thanked Johannes Stark for sending him "again an original photograph," which he used for a lecture. Stark's research in 1907 dealt with luminescence, a theme with which Sommerfeld had no prior familiarity. But he did not hesitate to approach it: "Because I am just lecturing on thermodynamic radiation," he thanked Stark, "luminescence should not be too far off. I presume, however, that a general theory is still beyond reach for a long time. In any case your experiments will make a considerable contribution."[22]

The readiness to address new subjects, even if these were poorly understood, combined with an awareness of what his students could appropriate, made Sommerfeld's

---

[20]"Um Erfahrungen zu sammeln, wollte er sich für zwei Stunden am Tag in meinem Labor umsehen. Stattdessen schlug ich ihm vor, nach dem Frühstück in das Café zu kommen, wo wir täglich physikalische Fragen diskutierten. Mit der ihm eigenen Gewissenhaftigkeit erschien Sommerfeld täglich ungefähr eine Stunde im Café Hofgarten, wo sich eine Art Physikerklub gebildet hatte, an dem auch Chemiker und Kristallographen teilnahmen und wo täglich über Fragen, die bei der Arbeit entstanden, diskutiert wurde. Sommerfeld fiel es manchmal nicht leicht, unseren Wortwechseln zu folgen, dafür übertraf sein Assistent Debye bald uns alle."

[21]Sommerfeld to Volterra, 3 January 1908, BANL (Volterra). "Nehmen Sie meinen verbindlichsten Dank für die Zusendung Ihrer schönen Vorlesungen. Die von Ihnen berührten Gegenstände haben mich früher vielfach beschäftigt, z.B. die Methode der Kelvin'schen Bilder, der multiformen Lösungen, sowie die allgemeine Theorie der Randwertaufgaben in meinem Encyklopädie-Artikel. Ich werde daher Ihre Vorlesungen genauer studiren. Zur Zeit bin ich allerdings mehr mit den realeren und spezielleren Fragen der Physik beschäftigt, wie mit den allgemeinen mathematischen Problemen, die Sie behandeln."

[22]Sommerfeld to Stark, 31 May 1907, SBPK (Stark Papers). Also in ASWB I. "Sie waren so gütig, mir neben Ihren Arbeiten wieder eine Originalphotographie zu schenken, wofür ich Ihnen herzlich danke. Da ich gerade ein kleines Colleg über thermodynamische Strahlung lese, sollte mir eigentlich auch die Luminiscenz-Strahlung nicht zu fern liegen. Aber mit einer allgemeinen Theorie derselben sieht es wohl auf lange böse aus. Ihre Versuche werden jedenfalls wesentlich dazu beitragen."

lectures very popular. Paul Peter Ewald, who had come to Munich in the fall of 1907 to continue his study of mathematics, was lured to Sommerfeld's hydrodynamics lecture in the summer of 1908 by a fellow student. "I think from the very first lecture this captured me entirely." Ewald recalled and later wrote,

> Sommerfeld had this beautiful way of not assuming anything. I had not done mechanics, not even point mechanics. And I went straightaway to hydrodynamics. And I had no idea of vectors, I had never heard of vectors. And Sommerfeld developed the whole vector algebra, scalar product, vector product, and so on, and that vector analysis, in conjunction with the hydrodynamic concepts.[23]

Memories like these, recollected by one of Sommerfeld's master pupils and close friend more than fifty years later, should not be taken at face value. Oral history interviews with Ewald, Debye, and other students tend to embellish the somewhat haphazard beginnings of the Munich nursery for theoretical physics. With regard to Sommerfeld's pedagogical talent, however, they all agree that it was rather exceptional. There is also evidence from Sommerfeld's early career that demonstrates his conscious efforts to positively engage the students with the subject. As early as 1895, when Sommerfeld presented his first mathematical lectures as Klein's assistant in Göttingen, he was very conscious about his performance as a teacher.[24] Even when he was unprepared, he impressed his students. "The joy about the subject matter transfers itself from the speaker to the audience," he explained once why his students enjoyed his teaching.[25] At the end of another semester he proudly reported to his mother that he received a "big applause" in his final lecture and that he completed this term with the "pleasant awareness that the people have learned a great deal and stayed interested in the subject matter from beginning to end."[26] On another opportunity, he wrote to his mother that his lecture was "magnificently wonderful" and how "simply excited" his students have been. "Thus my fame as an interesting lecturer is considerably consolidated."[27]

Sommerfeld's exuberant self-praise illustrates how proud he was about his teaching abilities. It is obvious in his review for the professorial chair position in Munich

---

[23] Interview with Ewald by George Uhlenbeck with Thomas S. Kuhn and Mrs. Ewald, 29 March and 8 May 1962, https://www.aip.org/history-programs/niels-bohr-library/oral-histories/4523-1, accessed 17 April 2020.

[24] Sommerfeld to his parents, 29 April 1895, Private Papers. "Das erste Lampenfieber überstanden, die Jungfernrede gehalten, 2 Stunden Colleg absolvirt. [...] Ich begann mit einer historischen Einleitung, war sehr fein präparirt; es machte mir viel Spass, den Zuhörern scheinbar auch."

[25] Sommerfeld to his parents, 19 December 1895, Private Papers. "In der letzten Woche habe ich stets ziemlich unpräparirt im Colleg vorgetragen. Ist aber doch ganz schön gegangen. Die Studenten haben mir meistens mit Freude zugeört. Das Vergnügen an der Sache überträgt sich von selbst von dem Redenden auf die Hörenden."

[26] Sommerfeld to his mother, 10 March 1896, Private Papers. "Mein Colleg habe ich mit Glanz am vorigen Freitag beendigt. Ich hatte noch einen besonderen Leckerbissen für die letzte Stunde verspart u. wurde mit riesigem Trampeln entlassen. Ich habe das angenehme Bewusstsein, dass die Leute einen grossen Posten bei mir gelernt haben u. dauernd für die Sache interessirt waren."

[27] Sommerfeld to his mother, 3 March 1897, Private Papers. "Mein Colleg ist grossartig schön gewesen. Die Studenten waren einfach begeistert. Mein Ruhm als interessanter Docent ist erheblich dadurch befestigt. Die Leute haben einen geradezu rührenden Fleiß entwickelt,"

that he did not exaggerate his reputation. "S. is characterized to us as a charming colleague and excellent teacher," the final report of the commission in Munich concluded
about Sommerfeld.[28] A few semesters in Munich spread his "fame as an interesting
teacher" elsewhere. In January 1908, Albert Einstein wrote to Sommerfeld that he
wished to be able to "sit in your lecture in order to accomplish my mathematical-
physical education."[29] Iris Runge, the daughter of the Göttingen mathematician Carl
Runge, went to Munich in 1910 for a short study sojourn. "Sommerfeld is wonderful, his way of tackling mechanics is just exciting to me," she marveled about
Sommerfeld's lecture course.[30] In 1911, Paul Ehrenfest confided to Sommerfeld that
he wished "for 2–3 years already" to study in Munich "under your personal supervision."[31] Ewald's recollection, therefore, only confirmed what had been often noted
as particularly remarkable about Sommerfeld: he was a charismatic teacher.

Supervising doctoral candidates was a key responsibility as a theoretical physics
professor. At the time of his arrival in Munich, Sommerfeld had had no experience
in this regard. Debye had only completed an engineering diploma during his first
years as Sommerfeld's assistant. In 1905, Sommerfeld had planned to send Debye to
Würzburg where he could pursue his doctorate in physics under Wien's supervision.[32]
The offer in Munich, however, made it clear that Debye would be Sommerfeld's own
doctoral student. The subject for Debye's dissertation emerged from a discussion
between Sommerfeld and Karl Schwarzschild about the diffraction of light. Applied
to tiny spherical bodies, the theory seemed appropriate for previously unexplained
phenomena. In 1901, Schwarzschild had calculated the pressure of light on perfectly
reflecting spheres and concluded that this pressure caused comet tails to deflect
from the sun.[33] Debye applied his theory also to water droplets and thus contributed
to the theory of the rainbow, as Sommerfeld remarked in his report on Debye's
dissertation.[34]

Another early doctoral candidate did not find Sommerfeld to be such an inspiring
mentor. Frederick W. Grover, who had come from the National Bureau of Standards
in Washington, D.C., to earn his doctoral degree under Sommerfeld's supervision,

[28]Report to the faculty, 20 July 1905. UAM, E–II–N, Sommerfeld. "S. wird uns als liebenswürdiger
Kollege und als ausgezeichneter Lehrer geschildert."

[29]Einstein to Sommerfeld, 14 January 1908, DM, NL 89, 007. Also in ASWB I. "Aber ich versichere
Ihnen, dass ich, wenn ich in München wäre und Zeit hätte, mich in Ihr Kolleg setzen würde, um
meine mathematisch-physikalischen Kenntnisse zu vervollständigen."

[30]Runge to her parents, 8 November 1910. Quoted in Tobies (2010, 72). "Sommerfeld ist herrlich,
seine Art die Mechanik anzupacken, begeistert mich geradezu [...]".

[31]Ehrenfest to Sommerfeld, 30 September 1911, DM, HS 1977–28/A,76. Also in ASWB I. "Ich
wünsche nämlich schon seit 2–3 Jahren (wie Herr Epstein weiß) nach München zu gehen, um unter
Ihrer persönlichen Leitung – neben vielem anderen – speciell dieses zu lernen: wie man eine Arbeit
die wirklichen Rechenaufwand erfordert zu Ende führt."

[32]Sommerfeld to Wien, 20 June 1905, DM, NL 56, 010. "Ich habe einen ganz genialen Assistenten,
der einmal bei Ihnen den Doctor machen soll."

[33]Schwarzschild to Sommerfeld, 15 June 1901, DM, HS 1977–28/A,318. Also in ASWB I.

[34]Sommerfeld's report (votum informativum) on Debye's doctoral work to the faculty, 23 July 1908,
UAM, OC–I–34p. Also in ASWB I.

was assigned a problem that involved precision measurements of eddy currents. Sommerfeld expected the problem to be explored both experimentally and theoretically. However, the experimental part could not be accomplished because the equipment in Sommerfeld's laboratory proved inadequate. Grover complained that he had never before worked under such primitive circumstances (Eckert 1999, 244).

By the summer of 1909, two other doctoral candidates completed their studies with Sommerfeld. Demetrios Hondros elaborated upon the theory of electromagnetic waves along wires, a theme with which Sommerfeld had a long-standing familiarity.[35] Ludwig Hopf presented a study of two hydrodynamical problems: one theoretical (ship waves), the other experimental (onset of turbulence). Sommerfeld praised Hopf for solving the theoretical part but considered him "not really experimentally skilled."[36] Hopf had performed his experiments in the Bavarian Academy rooms where Sommerfeld resided before his new institute was opened at the university. Sommerfeld moved to this new institute later that same year but never again asked his doctoral students to cope with experiments despite a modern workshop and a mechanic on-site. Grover's struggle with inadequate equipment and Hopf's poor performance as an experimental physicist must have been a warning sign for Sommerfeld to limit his pedagogical efforts henceforth to theory. Subsequent students in Sommerfeld's institute were seldom asked to perform experiments in the course of their doctoral research. Until the First World War, Sommerfeld had supervised nine other doctoral students. Only one of them (Wilhelm Hüter) performed experiments (Table A.2).

It is difficult to discern a theme among these doctoral dissertations. Many years later, Ewald recalled the situation when he asked Sommerfeld for a doctoral theme:

> Sommerfeld took a sheet out of the drawer on which were listed some ten or twelve topics suitable for doctoral theses. They ranged from hydrodynamics to improved calculations of the frequency dependence of the self-induction of solenoids and included various problems on the propagation of the waves in wireless telegraphy—all of them problems providing a sound training in solving partial differential equations with boundary values. At the end of the list stood the problem: "To find the optical properties of an anisotropic arrangement of isotropic resonators." Sommerfeld presented this last topic with the excuse that he should perhaps not have added it to the others since he had no definite idea of how to tackle it, whereas the other problems were solved by standard methods of which he had great experience. (Ewald 1962, 37)

Ewald's memory fifty years later fits with the diversity of Sommerfeld's own research themes and is corroborated by the topics of Sommerfeld's other doctoral students before World War I (Table A.2). From the vantage point of a prospective doctoral student, this openness was both an opportunity and a challenge to pursue self-imposed goals. Sommerfeld's lectures contained a plethora of specific problems with which a student could explore his skills. "And then what I had to do was make up the problems," Debye recalled of his duties as Sommerfeld's assistant. "Well, he had

---

[35]Sommerfeld's report (votum informativum) on Hondros's doctoral work to the faculty, 15 June 1909, UAM, OC–I–35p.

[36]Sommerfeld's report (votum informativum) on Hopf's doctoral work to the faculty, 5 July 1909, UAM, OC–I–35p. "Der Verf. ist experimentell nicht eigentlich geschickt." See also Eckert (2015).

in his course also one, it may have been two hours where the boys could handle their problems. They came there and delivered their problems and they were discussed." Before the First World War, these "seminars" were just occasions to discuss problems that emerged in Sommerfeld's lectures or were conceived by his assistant. "And what I had to do was nothing else than give this one or two hour seminar for the problems to be handled by the students. I did not even handle the problems. Sommerfeld wanted to do that. I had only to make the problems."[37] Sommerfeld would thus become aware if a student displayed enough talent and skill for a doctoral thesis. By 1909, this practice became extended by another proving ground, as Ewald recalled:

> I didn't like the idea that Debye and Hondros and Hopf and Hörschelmann—all these elderly students—talked about things I didn't understand at all. And so I said to Hondros, I urged very strongly: "Couldn't we get together to talk over things, so that we young people could learn a bit quicker what the problems of real actual interest were." Hondros agreed that would be a great plan. He talked to Debye, and Debye talked to Sommerfeld. Sommerfeld said: "Oh, that's fine," and Debye said yes. Sommerfeld said: "I will not be there, so you are quite among yourselves." Which I think was a very wise decision. But he bought a box of cigars to be put on the table and to be smoked during the colloquium, which was his gift to the colloquium.[38]

Ewald was not the only one who claimed to have sparked the creation of the "Physical Wednesday Colloquium," as it became called.[39] Nor was it established in just the manner he recollected. Debye was assigned the responsibilty of creating the colloquium, which was intended for the "elderly students"[40] whom Ewald found so difficult to understand, not the students in their early semeseters. Furthermore, Sommerfeld did not abstain from attending and occasionally presented subjects that he considered interesting himself.[41] Nevertheless, it was less formal than the established big physics colloquium, which was considered tradition in most German physics institutes since the late nineteenth century. Before World War I, Sommerfeld's seminar and the Wednesday Colloquium were "approximately the same thing," Debye

---

[37] Interview with Debye by Kuhn and Uhlenbeck, 3 May 1962, https://www.aip.org/history-programs/niels-bohr-library/oral-histories/4568-1, accessed 17 April 2020.

[38] Interview with Ewald by Uhlenbeck with Kuhn, and Ella Ewald, 29 March and 8 May 1962, https://www.aip.org/history-programs/niels-bohr-library/oral-histories/4523-1, accessed 17 April 2020.

[39] Peter P. Koch, then Privatdozent in Röntgen's institute, wrote to Sommerfeld on 6 August 1944, ULM, Sommerfeld Papers: "Die höchste Spitze in der Aus- und Fortbildung des Physikers sah ich von jeher im Physikalischen Kolloquium. Das war schon so in meiner Münchner Zeit, in der ich als junger Privatdozent zusammen mit Debye und Wagner das 'bonzenfreie' Kolloquium gründete, aus dem dann später das berühmte Sommerfeldkolloquium erwuchs."

[40] Richard Hertwig (then rector of Munich University) to Sommerfeld, 3 November 1910, DM, NL 89, 030. "Nachdem Ihr Assistent Dr. Debye als Privatdozent an unserer Universität zugelassen ist, besteht (entsprechend einem Senatsbeschluß vom 19. Dezember 1903) kein Bedenken dagegen, dass in der Zeit von 6–8 Uhr im Hörsaal Nr. 122 unter seiner Leitung ein Colloquium mit älteren Studierenden abgehalten wird."

[41] Physikalisches Mittwoch-Colloquium, DM, 1997–5115.

recalled many years later. "The main point was that there was talk about a subject and then there was discussion about it."[42]

It was also an event that fostered a sense of togetherness. On the last Wednesday of each semester, the colloquium used to go bowling in the basement of a Munich beer hall. Sommerfeld further encouraged his advanced students' enthusiasm by inviting them to his home or on skiing trips in the Alps. The recollections of his disciples become enthusiastic with memories of such social events. "Now I understand why your disciples are so fond of you," wrote Einstein to Sommerfeld in September 1909 after they had first met personally. "Such a beautiful relationship between professor and student stands out as unique. I am determined to take you as my role model."[43]

The study of theoretical physics in Sommerfeld's circle was definitively not a solitary effort. Nor was it limited to theory, as illustrated by Ewald's demonstration of glider flight in the colloquium on November 10, 1909 and a report by Röntgen's doctoral student Walter Friedrich about his experiments with polarized X-rays on November 15, 1911.[44] Although Sommerfeld no longer considered it necessary for his own doctoral students to perform experiments, he considered it desirable to keep close contact with experimental physics. He admired those who managed to perform both excellent theoretical and experimental work as true physicists. His assistant, Debye, came close to this ideal. Sommerfeld recommended Debye as Einstein's successor at the university in Zurich, with the explicit praise of Debye's "practical intuition and experimental skill to which I cannot compare myself."[45] In 1911, when the university finally approved the second assistant position in Sommerfeld's institute (the first one was on the payroll of the Bavarian Academy of Science because of the connection with the state's mathematical-physical instrument collection), Sommerfeld hired Friedrich (Sommerfeld 1926). Sommerfeld disposed of means and personnel that in retrospect seem unusual for an institute dedicated to theoretical physics: a new doctoral student, Wilhelm Lenz, as theoretical assistant, experimental physicist, Friedrich, as second assistant, a mechanic, and a basement equipped for experimental work. His sudden embrace of experimental research apparently also provoked a departmental competition. "Röntgen wants Friedrich for himself," Sommerfeld wrote to his wife, Johanna, shortly after he had offered Friedrich the position as his assistant.

---

[42]Interview with Debye by Kuhn and Uhlenbeck, 3 May 1962, https://www.aip.org/history-programs/niels-bohr-library/oral-histories/4568-1, accessed 17 April 2020.

[43]Einstein to Sommerfeld, 29 September 1909, NMAH, MSS 122A, Einstein Papers. Also in ASWB I. "Ich begreife es jetzt, dass Ihre Schüler Sie so gern haben! Ein so schönes Verhältnis zwischen Professor und Studenten steht wohl einzig da. Ich will mir Sie ganz zum Vorbild nehmen."

[44]Physikalisches Mittwoch-Colloquium, DM, 1997–5115.

[45]Recommendation for Debye, 30 March 1911, UAZ, SA, U 110b.2. This passage was communicated in a letter of Hans Schinz to the "Direktion des Erziehungswesens des Kanton Zürich." "Ich schätze die absolute Zuverlässigkeit und Ehrlichkeit seines Charakters ebenso sehr wie seine Intelligenz, die ich mir oft überlegen fühle und seinen praktischen Blick und seine experimentelle Geschicklichkeit, in der ich mich ihm nicht vergleichen kann."

R. is stupid enough to demand an immediate decision within two hours. Friedrich declines. I am very glad about it, not only because I need Friedrich but also because one does not like to lose in a showdown. It does no harm that R. will have the opposite feelings.[46]

However, the showdown did not result in a feud. With Röntgen "everything is OK after it was unpleasant for a while," Sommerfeld wrote a few days later. "He is really an excellent man."[47] Sommerfeld's respect for the great experimentalist and his taking care of university affairs when Röntgen was sick apparently restored fleeting resentment caused by this and other incidences.[48]

References

Eckert M (1997) Mathematik auf Abwegen: Ferdinand Lindemann und die Elektronentheorie. Centaurus 39:121–140

Eckert M (1999) Mathematics, Experiments, and Theoretical Physics: The Early Days of the Sommerfeld School. Phys Perspect 1:238–252

Eckert M (2015) Fluid Mechanics in Sommerfeld's School. Ann Rev Fluid Mech 47:1–20

Eckert M, Pricha W et al (1984) Geheimrat Sommerfeld-Theoretischer Physiker. Deutsches Museum, Munich

Ewald PP (1962) Arnold Sommerfeld als Mensch, Lehrer und Freund. Rede, gehalten zur Feier der 100sten Wiederkehr seiner Geburt. In: Bopp F, Kleinpoppen H (eds) Physics of the One- and Two-Electron Atoms. North Holland, Amsterdam, pp 8–16

Fölsing A (1995) Wilhelm Conrad Röntgen: Aufbruch ins Innere der Materie. Hanser, Munich

Graetz L (26/27 November 1926) Physik der letzten hundert Jahre: Ihre Pflege in München. In: Münchner Neueste Nachrichten. Sonderbeilage

Ioffe AF (1967) Begegnungen mit Physikern. Pfalz-Verlag, Basel

Jungnickel C, McCormmach R (1986) Intellectual Mastery of Nature: Theoretical Physics from Ohm to Einstein, vol 2. The University of Chicago Press, Chicago

Litten F (1993) Die Korn-Röntgen-Affäre. Kultur Technik 17(4):42–49

Sommerfeld A (1926) Das Institut für theoretische Physik. In: von Müller KA (ed) Die wissenschaftlichen Anstalten der Ludwig-Maximilians-Universität zu München: Chronik zur Jahrhundertfeier im Auftrag des akademischen Senats. Oldenbourg, Munich, 290–6292

Tobies R (2010) "Morgen möchte ich wieder 100 herrliche Sachen ausrechnen" : Iris Runge bei Osram und Telefunken. Steiner, Stuttgart

Zehnder L (ed) (1935) W. C. Röntgen: Briefe an L. Zehnder. Rascher, Zurich

---

[46] Sommerfeld to his wife, undated [20 July 1912]. Private Papers. "Röntgen will Friedrich für sich haben. [...] Da R. so töricht ist, sofortige Entscheidung zu verlangen, in 2 Stunden, lehnt Friedrich ab. Ich freue mich sehr darüber, nicht nur weil ich F. brauche, sondern auch weil man bei einer Kraftprobe nicht gerne unterliegt. Daß R. die umgekehrten Empfindungen haben wird, schadet nichts."

[47] Sommerfeld to his wife, 26 July 1912. Private Papers. "Mit Röntgen ist alles in Ordnung, nachdem es vorübergehend unangenehm war. Er ist wirklich ein famoser Mann."

[48] Röntgen to Sommerfeld, 14 December 1912, DM, HS 1977–28/A,288.

Chapter 4
X-rays and Quanta, 1911–1913

Abstract As a colleague of Wilhelm Conrad Röntgen, Arnold Sommerfeld perceived the still unexplained nature of X-rays as a major challenge. From this context emerged Sommerfeld's "h-hypothesis," an attempt to explain elementary atomic processes in terms of Planck's quantum constant. In the context of this effort, Sommerfeld was invited to the first Solvay Conference in 1911. An unintended consequence was the discovery at Sommerfeld's institute of X-ray diffraction by crystals. At the suggestion of Sommerfeld's *Privatdozent* Max von Laue, the experiment was conducted by Röntgen's doctoral students Walter Friedrich and Paul Knipping. Although the "h-hypothesis" was doomed to fail and the "Laue experiment" was based on false concepts, both investigations added to the reputation of Sommerfeld's institute as an aspiring center at the forefront of atomic physics.

Keywords X-rays · Bremsstrahlen · h-hypothesis · Solvay conference · X-ray diffraction by crystals · Max von Laue · Walter Friedrich · Paul Knipping

More was at stake with Sommerfeld's interest in experimental research on X-rays. By the time he had lured Röntgen's doctoral student Friedrich away to become his own experimental assistant, Sommerfeld had been researching X-rays for a dozen years already. In 1900, Sommerfeld attempted to explain Hermanus Haga and Cornelis Wind's experiments on X-ray diffraction behind V-shaped slits. Based on the widespread assumption that X-rays consist of pulse-like disturbances of electromagnetic ether (Wheaton 1983, Chap. 2), Sommerfeld derived that the "width" of an X-ray pulse was about 1 Angstrom. This is "the order of magnitude of molecules," Sommerfeld concluded of the physical consequences of his theory. "At this size it is completely plausible why X-rays are not absorbed or diffracted."[1] However, both the experimental evidence and the theoretical elaboration left room for controversial debates. Already in 1905, Sommerfeld found it a disgrace that the nature of X-rays

---

[1]Sommerfeld to Schwarzschild, 24 September 1900, SUB, Schwarzschild 743. "[…] das ist die Grössenordnung der Moleküle. Bei dieser Kleinheit versteht man vollkommen, warum Röntgenstrahlen nicht absorbirt u. gebrochen werden."

M. Eckert, *Establishing Quantum Physics in Munich*,
SpringerBriefs in History of Science and Technology,
https://doi.org/10.1007/978-3-030-62034-9_4

was not known.[2] When he became Röntgen's colleague a year later, he increasingly perceived this shame as a challenge.

Within few years, it had been discovered that X-rays come in two varieties. One sort of X-ray was independent of the anode material. In 1909, Sommerfeld recognized that electromagnetic radiation was caused by electrons at their impact with the anode of an X-ray tube during their deceleration inside the anode material. Consequently, he named this kind of X-ray *Bremsstrahlung* (Sommerfeld 1909, 1910). The second variety had the characteristic of fluorescent radiation. The bremsstrahlung was polarized, and it displayed an angular distribution of intensity dependent on the energy of the electrons at the impact in the anode. Fluorescent X-rays, however, were unpolarized and characteristic for the material of the anode. Further clues about the nature of X-rays were added by experiments performed in Wien's laboratory in Würzburg, which established a relation between the energy of X-rays to the energy of electrons that created them upon impact in the anode. Sommerfeld considered both Wien's experiments and recent experiments in Röntgen's laboratory compatible with his bremsstrahlen theory. However, the agreement between theory and experiment was deficient. On the experimental side, it remained undetermined how much of the electrons' energy at impact went into the production of unpolarized fluorescent X-rays and how much into the production of bremsstrahlen. The latter became Friedrich's first task when he began work in Sommerfeld's basement laboratory. "I have recruited Friedrich for the elaboration of my X-ray problem," Sommerfeld wrote to his wife in July 1911.[3] Several months later, he informed Robert Wichard Pohl, an experimental physicist who had recently repeated the Haga-Wind-experiments with greater precision, that Friedrich had begun measuring the "polarized energy" of X-rays.[4] Friedrich was perfectly prepared for this work because he had performed similar experiments in Röntgen's laboratory during his doctoral work (Friedrich 1912).

On the theoretical side, Sommerfeld's bremsstrahlen theory could not be elaborated without further assumptions about the impact of electrons in the anti-cathode of an X-ray tube. At this point, Sommerfeld closed his theory using a quantum hypothesis: he linked the time required to stop an electron, τ, and the energy released in this process, E, to Planck's quantum of action, h, via $\tau E = h$. This so-called h-hypothesis predicted, for example, a ratio of about 6,000 between the electron energy and the energy of the polarized X-rays under the circumstances of Wien's earlier experiments, but the comparison was based on estimates concerning the relation between polarized and unpolarized X-rays. He employed the same hypothesis to explain the nature of γ-rays as a by-product of the emission of β-rays, which he assumed would

[2]Sommerfeld to Wien, 13 May 1905, DM, NL 56, 010. Also in ASWB I. "Es ist eigentlich eine Schmach, daß man 10 Jahre nach der Röntgen'schen Entdeckung immer noch nicht weiß, was in den Röntgenstr. eigentl. los ist."

[3]Sommerfeld to his wife, 22 July 1911, Private Papers. "Friedrich habe ich an Stelle von Herweg zur Bearbeitung meines Röntgenproblems angeworben."

[4]Sommerfeld to Pohl, 1 May 1912. Private papers of Robert Wichard Pohl. I am grateful to Pohl's son, Robert Pohl, for a copy of this letter. "Herr Dr. Friedrich hat bei mir Versuche begonnen, die dahin gehen, das entsprechende für die polarisierte Energie zu machen." On Pohl and Walter's experiments, see (Wheaton 1983, Chap. 6).

be accelerated within atoms in a sort of inverted process, just as X-rays are caused by decelerated electrons (Sommerfeld 1911a, 24–25).

Sommerfeld's h-hypothesis implied that the greater or smaller hardness of X-rays resulted from a smaller or larger duration of the deceleration process in the anti-cathode of an X-ray tube. He concluded that "molecular processes" are ruled by laws that appear strange compared to analogous "ballistic experiences," where it would be absurd to assume that fast bullets are stopped within a shorter time than slow bullets when they hit the target. Sommerfeld held great expectations in this hypothesis and believed it would be applicable to other "molecular" phenomena, like the photoelectric effect. One should not attempt to explain h from electromagnetic or mechanical models, Sommerfeld concluded in a public presentation at an annual meeting of natural scientists in the fall of 1911, but take h as an axiomatic foundation of atomic theory (Sommerfeld 1911b).

In October 1911, Sommerfeld spread the same gospel at the first Solvay Conference, which was dedicated to the theory of radiation and quanta (Sommerfeld 1912). Although his "h-hypothesis" received criticism, the quantum effort at Munich was met with curiosity. However, when Lorentz later asked whether Sommerfeld's great expectations in his h-hypothesis had been fulfilled, there was still no experimental evidence forthcoming. Sommerfeld responded with a verse by Johann von Goethe, saying that it is like planting roses without knowing whether they will bloom. "The experiments with X-rays, which are supposed to tell me something about the likelihood of the blooming, are not yet ready."[5]

Once more, an unexpected twist changed whatever Sommerfeld may have planned if his approach with the h-hypothesis had blossomed. The "experiments with X-rays" that Friedrich was preparing in the basement of Sommerfeld's institute were interrupted when von Laue[6] suggested another X-ray experiment: to bombard a crystal with X-rays so that it emits fluorescent radiation. He expected that these secondary X-rays would interfere with one another and yield a regular diffraction pattern because they were emitted by the regularly spaced atoms in the crystal. Sommerfeld apparently refused to excuse Friedrich from his original experimental plans so von Laue asked another experimental physicist from Röntgen's laboratory, Paul Knipping, to perform the experiment. Why and how Friedrich and Knipping followed von Laue's suggestion became subject of controversial recollections (Forman 1969; Ewald 1969a; Eckert 2012). But when they actually discovered a diffraction pattern, Sommerfeld threw out his earlier experimental project and focused on the exploration of X-ray interference on crystals. Although the discovery was explained in a different manner than what von Laue had initially expected, "von Laue's discovery" was celebrated as an outstanding triumph of Sommerfeld's institute. "Von Laue's discovery is very beautiful," Debye reacted when he learned about it in May 1912.

---

[5]Sommerfeld to Lorentz, 25 February 1912. RANH (Lorentz Papers). "Da hilft nun weiter kein Bemühn, Sinds Rosen nun sie werden blühn.' Die Versuche mit Röntgenstrahlen, die mich über die Wahrscheinlichkeit des Blühens näher unterrichten sollten sind noch nicht fertig." See also Eckert (2015a, b).

[6]Laue's father was ennobled in 1913 so that Max Laue became Max von Laue. In order to avoid confusion, we leave the "von" in Laue's name throughout this book

He considered von Laue's discovery to be accidental to some extent and regarded Sommerfeld's long-standing interest in X-rays as the true cause for the discovery. "Therefore, I feel that I should first of all congratulate you to this success," he wrote to Sommerfeld.[7]

Although the circumstances of the discovery caused internal trouble,[8] Sommerfeld praised it like few other of his institute's achievements. The "wonderful interference photographs" made at von Laue's request "keep us all on tenterhooks," Sommerfeld wrote to a colleague.[9] X-ray diffraction by crystals also became a theme at the second Solvay Conference in 1913, dedicated to "The Structure of Matter." In 1914, von Laue was awarded the Nobel Prize. The contribution to understanding crystals' interference of X-rays not only enabled the growth of an entirely new field of X-ray crystallography, with ramifications in solid state physics and material science, it also gave rise to X-ray spectroscopy, and thus opened a new avenue for research on the structure of atoms. As in the years before "von Laue's discovery" when X-rays had been Sommerfeld's own research field, "X-rays and crystals" now became an important part of his institute's research program. Ewald, whose dissertation on crystal optics inspired von Laue's idea, dedicated his entire career to this program, first succeeding Friedrich as Sommerfeld's assistant, then as professor of theoretical physics at the Technical University in Stuttgart (now the University of Stuttgart), and later in exile and as president of the International Union of Crystallography (Bethe and Hildebrandt 1988). Even in 1926, when Sommerfeld had good reasons to be proud of the merits of his nursery in quantum theory, he considered the discovery of X-ray diffraction on crystals as the "most important scientific accomplishment in the history of the institute" (Sommerfeld 1926, 291).[10]

What about Sommerfeld's original plan, the experimental test of the h-hypothesis, for which he had lured Friedrich away from Röntgen's institute? In 1913, Sommerfeld and Debye expanded upon a theory of the photoelectric effect as a last attempt

---

[7]Debye to Sommerfeld, 13 May 1912, DM, HS 1977–28/A,61. "Die Laue'sche Entdeckung ist sehr schön [...] Zwar soll man bei solchen Sachen im allgemeinen Verdienst und Zufall nicht gegeneinander abwägen, aber eines muss ich sagen. Hättest Du dich nicht schon lange für Röntgenstrahlen interessiert, hättest Du nicht die Mittel Deines Instituts in liberalster Weise zur Verfügung gestellt und nicht jedem immer freien Einblick in Deine Gedanken gewährt, es wäre Laue nicht eingefallen und er hätte vor allem nicht die praktisch geschulten Mitarbeiter gefunden, welche unerlässlich zum Gelingen waren. So kommt es, dass ich das Gefühl habe, Dich zuerst zu diesem Erfolg gratulieren zu sollen."

[8]Von Laue felt isolated among the Munich physicists. See von Laue to Sommerfeld, 3 August 1920, DM, HS 1977–28/A,197. Also in ASWB II. "Warum haben Sie mich ausgeschlossen, als Sie mit Friedrich und Knipping und den anderen jüngeren Fachgenossen die Entdeckung der Röntgenstrahlinterferenzen feierten?".

[9]Sommerfeld to Alfred Kleiner, 13 May 1912, AETH, HS 412:2. Also in ASWB I. "Laue wird jedenfalls mit Freuden die Gelegenheit wahrnehmen Ihnen die wundervollen Interferenz-Aufnahmen mit Röntgenstrahlen an Krystallen zu zeigen, die jetzt hier auf seine Veranlassung gemacht werden u. die uns alle in Atem halten."

[10]"Das wichtigste wissenschaftliche Ereignis in der Geschichte des Instituts war die Lauesche Entdeckung im Jahre 1912, bei der die Herren Dr. Friedrich als Institutsassistent und Dr. Knipping mitwirkten."

to make the h-hypothesis plausible. They assumed that an atom of a metal collects light over a period of time ("accumulation time") until the energy is large enough to emit an electron. Although they were able to derive Einstein's law for the photo-electric effect, the theory amounted to unrealistic conclusions about the accumula-tion process (Wheaton 1983, 180–189). Depending on the wavelength of the light with which photoelectrons were emitted, the accumulation time varied widely. X-ray wavelengths amounted to years! Ten years later, when Sommerfeld published the third edition of *Atombau und Spektrallinien*, he mentioned this absurd result as an example for the futile attempts to marry the quantum theoretical with classical conceptions (Sommerfeld 1922, 54).

But Sommerfeld's h-hypothesis was not the only contribution to the early quantum theory originating in Munich. Debye's last feat in Munich (before he was called to Zurich as Einstein's successor) was a novel derivation of the law of black-body radiation by quantizing the vibrational modes of radiation in a cavity, rendering unnecessary Planck's disputed oscillators (Debye 1910; Hermann 1969, 125–126; Kuhn 1978, 209–210, 308).[11] A year later, Debye succeeded with a similar approach to explain the thermal behavior of lattice vibrations, almost simultaneously with a similar attempt by Born and Theodore von Kármán in Göttingen. The analysis of vibrational modes was therefore regarded as another quantum playground-in Munich and elsewhere (Eckert et al. 1992, 33–34).

With Debye's quantum theory of the specific heat of solids, Sommerfeld's h-hypothesis, and other efforts to introduce Planck's h as the means to clarify heretofore unexplained phenomena, quantum physics became an attractive topic for special lectures and conferences. In April 1912, Hilbert invited Sommerfeld to lecture on quantum theory in Göttingen at the end of the forthcoming summer semester. Instead, Sommerfeld suggested inviting Planck and Einstein "who have dealt longer and more profoundly than I with quantum theory and are much brighter fellows than myself," but he ultimately accepted the invitation with the qualification that his quantum views were "still very embryonic and little accomplished."[12] He intended to focus on Debye's recent quantum theory of specific heats.[13] At the same time, he assigned the "method of vibrational modes," as Debye's approach was called, to Alfred Landé as a theme for a doctoral thesis (Landé 1914).[14] If Debye had accomplished quantizing the

---

[11]Debye to Sommerfeld, 2 March 1910. DMA, HS 1977–28/A,61. Also in ASWB I.

[12]Sommerfeld to Hilbert, 10 April 1912, SUB, Cod. Ms. Hilbert 379A. "Dass Sie an mich denken, wo Sie doch Planck und Einstein haben könnten, die sich viel länger und gründlicher mit der Quantentheorie befasst haben und überhaupt viel gescheitere Kerle sind wie ich, dies führe ich lediglich auf Ihre freundschaftliche Gesinnung gegen mich zurück. Überlegen Sie es nur nochmals, ob Sie nicht doch lieber einen von diesen nehmen. Ich würde mir natürlich sehr viel Mühe geben und Ihnen nicht nur meinen eigenen Standpunkt gegen diese Fragen schildern, der noch sehr embryonal und wenig fertig ist, sondern auch die anderen Auffassungen gebührend zu Worte kommen lassen." On Hilbert's interest in quantum physics during these years, see (Schirrmacher 2019).

[13]Sommerfeld to Schwarzschild, undated [early June 1912], SUB, Schwarzschild 743. Also in ASWB I. "Ich soll 29. Juli und 2. August in Göttingen über Quanten vortragen und werde vor allem die specif. Wärmen auf's Korn nehmen."

[14]Interview with Landé by Kuhn and Heilbron, 5 March 1962, https://www.aip.org/history-programs/niels-bohr-library/oral-histories/4728-1, accessed 17 April 2020.

electromagnetic radiation in a cavity and the lattice vibrations of a solid, why not also analyze the vibrational modes of electrons in a solid this way in order to account for optical dispersion?[15] With a similar motivation, Sommerfeld asked Wilhelm Lenz, who had succeeded Debye as his assistant, to analyze the vibrational modes of a gas in a container. When Hilbert invited him to Göttingen for a conference on the kinetic theory of gases in April 1913 (*Gaswoche*),[16] Sommerfeld presented the results of Lenz's effort with the perspective that the quantum deviations predicted by this theory could be pertinent for the electron theory of metals, which the classical electron gas model failed to account for in a variety of phenomena. However, as he confided to Hilbert half a year later, he did not consider it "a satisfying accomplishment."[17]

By 1913, when the efforts of Lenz and Landé proved futile to account for gas degeneration and optical dispersion, and Sommerfeld's and Debye's *h*-hypothesis failed with the photoelectric effect, the Munich nursery could hardly claim to be a successful quantum school. Nevertheless, these attempts demonstrated that Sommerfeld did not shy away from big challenges. The X-ray diffraction by crystals illustrates how great discoveries were made, even when the original motivations and final results displayed unexpected twists and turns. The quantum failures certainly did not reduce the attractiveness of Sommerfeld's institute. Leon Brillouin, for example, who later rose to prominence with contributions to semi-classical quantum approaches and to the quantum theory of solids, was motivated to study with Sommerfeld by his father, Marcel Brillouin. The elder Brillouin had attended the first Solvay Conference. "They were all very much impressed with this Sommerfeld discussion," he recalled of his conversation with his father and other French participants, like Paul Langevin and Henri Poincaré, in the wake of the Solvay Conference. "And so I decided to go to Munich. This connection between my father and Sommerfeld was the reason why I decided to go to Munich."[18]

Hilbert also succumbed to the charm of the Munich nursery; he developed a keen interest in quantum physics and asked Sommerfeld to send him "physics assistants" to stay abreast of contemporary developments. Both of Sommerfeld's disciples, Ewald and Landé, spent a year with Hilbert in this capacity before the First World War.[19] Sommerfeld always kept close contact with his Göttingen colleagues and friends. He regarded himself as a scientific offspring of Göttingen rather than Königsberg. To satisfy Hilbert's request for "physics assistants," therefore, was more than col-

---

[15]Sommerfeld's report to the faculty about Landé's thesis, 28 April 1914. UAM, OC I 40 p. "Das Interesse an der Fragestellung lag für mich darin, zu wissen, ob man die Gesamtenergie [...] nach Quanten zu verteilen hat, oder ob sich etwa die Ätherenergie und die Elektronenenergie einzeln quantenhaft verhalten."

[16]The *Gaswoche* took place 21–26 April 1913. See Schirrmacher (2019).

[17]Sommerfeld to Hilbert, 14 October 1913, SUB, Cod. Ms. Hilbert 379 "Mein Göttinger Vortrag von der Gasconferenz, mit dem ich bezüglich der einatomigen Gase zu keinem befriedigendem Abschluss gekommen bin, ist nun im Druck."

[18]Interview with Brillouin by Ewald, Uhlenbeck, Kuhn and Ella Ewald, 29 March 1962, https://www.aip.org/history-programs/niels-bohr-library/oral-histories/4538-1, accessed 17 April 2020.

[19]See Schirrmacher (2019).

legial kindness. Henceforth, the Munich "nursery of theoretical physics" and the mathematical "Mecca" at Göttingen would become even more closely affiliated.

With his disciples in responsible positions elsewhere, Sommerfeld's school started to form a network. The network's outreach remained unobtrusive, despite the presence of Debye in Zurich and from 1912 in Utrecht, von Laue (as Debye's successor) in Zurich, and Ewald and Landé in Göttingen. However, within the small world of pre-war theoretical physics, even a small network could exert some influence. When Debye began to teach theoretical physics in Utrecht, he explicitly attempted to establish a "school" by following Sommerfeld's model. "So I will try in good cheer to cross Dutch thoughtfulness and endurance with German daringness toward a new joyful race."[20] Debye's first measures along this path included special lectures on quantum topics and the establishment of a "very casual colloquium" to which he invited physicists from Leiden, where Ehrenfest had just been called to succeed Lorentz. He hoped that "the smallness of the Netherlands" would facilitate collaboration between both universities.[21]

The same sort of networking happened in Munich where both the Wednesday Colloquium and the traditional Monday or "Sohncke" colloquium offered occasions to meet colleagues from the Technical University in Munich. At such opportunities, recent papers of common interest were discussed and often resulted in invitations of their authors so that the Munich physicists would receive first-hand knowledge about current research performed elsewhere. Epstein, for example, who had come to Munich in 1910 to study with Sommerfeld, reported in these colloquia on "Nernst's heat theorem" (28 June 1911), "Magnetons" (6 March 1911), and the "Recent work of Lebedev and his disciples" (17 July 1912) with whom he had studied in Moscow before his arrival in Munich. Sommerfeld was acquainted personally or through correspondence with the leading experts of a host of specialties.

References

Bethe HA, Hildebrandt G (1988) Paul Peter Ewald: 23 January 1888–22 August 1985. Biogr Mem Fellows R Soc 34:134–176
Debye P (1910) Der Wahrscheinlichkeitsbegriff in der Theorie der Strahlung. Annalen der Physik 33:142–1434
Eckert M (2012) Disputed Discovery: The Beginnings of X-ray Diffraction in Crystals in 1912 and Its Repercussions. Acta Crystallogr Sect A 68(1):30–39

---

[20]Debye to Sommerfeld, 29 March 1912, DM, HS 1977–28/A,61. "So will ich denn mit frischem Mut versuchen die Holländische Bedachtsamkeit und Ausdauer mit dem Deutschen Wagemut zu einer neuen, erfreulichen Rasse zu kreuzen."

[21]Debye to Sommerfeld, 3 November 1912, DM, HS 1977–28/A,61. "Für das zweite Semester habe ich angekündigt 1) Thermodynamik 2) Nernst'sches Theorem und Quantentheorie. Um einen Physikerkreis zu constituieren habe ich mit Julius zusammen ein ganz zwangloses Colloquium errichtet [...] Die Kleinheit von Holland liesz sich auch sehr schön folgendermaszen benutzen [...]."

Eckert M (2015a) From aether impulse to QED: Sommerfeld and the Bremsstrahlen theory. Stud Hist Philos Mod Phys 51:9–22

Eckert M (2015b) From X-rays to the h-hypothesis: Sommerfeld and the early quantum theory 1909–1913. Eur Phys J Spec Top 224:2057–2073

Eckert M, Schubert H, Torkar G (1992) The Roots of Solid- State Physics before Quantum Mechanics. In: Hoddeson L et al (eds) Out of the Crystal Maze. Chapters from the History of Solid-State Physics, Oxford University Press, New York, pp 3–87

Ewald PP (1969) The Myth of the Myths: Comments on P. Forman's Paper on "The Discovery of the Diffraction of X-Rays in Crystals". AHES 6:72–681

Forman P (1969) The Discovery of the Diffraction of X-Rays by Crystals; A Critique of the Myths. AHES 6:38–71

Friedrich W (1912) Räumliche Intensitätsverteilung der X-Strahlen, die von einer Platinantikathode ausgehen. Annalen der Physik 39:377–430

Hermann A (1969) Frühgeschichte der Quantentheorie. Physik-Verlag, Mosbach

Kuhn TS (1978) Black-Body Theory and the Quantum Discontinuity, 1894–1912. The University of Chicago Press, Chicago

Landé A (1914) Zur Methode der Eigenschwingungen der Quantentheorie. Phil. Diss. v. 29. Mai 1914. Universität München, Göttingen

Schirrmacher A (2019) Establishing Quantum Physics in Göttingen: David Hilbert, Max Born, and Peter Debye in Context, 1900–1926. Springer Briefs in History of Science and Technology, Springer Nature Switzerland AG, Cham

Sommerfeld A (1909) Über die Verteilung der Intensität bei der Emission von Röntgenstrahlen. Physikalische Zeitschrift 10:969–976

Sommerfeld A (1910) Über die Verteilung der Intensität bei der Emission von Röntgenstrahlen. Physikalische Zeitschrift 11:99–101

Sommerfeld A (1911a) Das Plancksche Wirkungsquantum und seine allgemeine Bedeutung für die Molekularphysik. Verhandlungen der Deutschen Physikalischen Gesellschaft 13:1074–1093

Sommerfeld A (1911b) Über die Struktur der Γ-Strahlen. In: Sitzungsberichte der mathematisch-physikalischen Klasse der K. B. Akademie der Wissenschaften zu München, pp. 1–60. Vorgetragen in der Sitzung am 7. Januar 1911

Sommerfeld A (1912) Sur l'application de la théorie de l'élément d' action aux phénomènes moléculaires non périodiques. Gauthiers-Villars, Paris

Sommerfeld A (1922) Atombau und Spektrallinien, 3rd edn. Vieweg, Braunschweig

Sommerfeld A (1926) Das Institut für theoretische Physik. In: Alexander K (ed) Die wissenschaftlichen Anstalten der Ludwig-Maximilians-Universität zu München: Chronik zur Jahrhundertfeier im Auftrag des akademischen Senats. von Müller, Munich, Oldenbourg, pp 290–292

Wheaton B (1983) The Tiger and the Shark: Empirical Roots of Wave-particle Dualism. Cambridge University Press, Cambridge

Chapter 5
Extending Bohr's Model, 1914–1919

Abstract Arnold Sommerfeld extended Niels Bohr's quantum model of the atom in order to explain the splitting of spectral lines in magnetic (Zeeman effect) and electric fields (Stark effect). Elaborating the atomic theory of spectra became a concerted effort that involved Walther Kossel (X-ray spectra), Paul Epstein and Karl Schwarzschild (Stark effect), Wilhelm Lenz (fine structure constant), Adalbert Rubinowicz (intensity and selection rules), and a close collaboration with Friedrich Paschen (optical spectroscopic precision measurements) and Manne Siegbahn (X-ray spectroscopy). Within a few years of the publication of Bohr's model in 1913, and despite World War I, atomic and quantum theory was transformed into a budding research field. In 1919, Sommerfeld reviewed these achievements in *Atombau und Spektrallinien*, a book that was soon regarded as the "bible" of atomic physics.

Keywords Niels Bohr · Bohr's atomic model · Bohr-Sommerfeld atom · Zeeman effect · Stark effect · Spectral lines · Fine structure · Walther Kossel · Paul Epstein · Karl Schwarzschild · Wilhelm Lenz · Adalbert Rubinowicz · Friedrich Paschen · Manne Siegbahn

It would be erroneous to conclude from these examples that Sommerfeld focused on quantum theory before World War I. Although his performance at the first Solvay Conference and Debye's quantum papers on black-body radiation and the specific heat of solids suggest that Sommerfeld was an early proponent of quantum theory, extensive research conducted at Sommerfeld's school before the First World War belie this impression. Planck's quantum constant h is absent in Sommerfeld's research papers prior to 1911, and only four out of sixteen articles published during the four years before the war may be categorized as early quantum theory.[1] Among Sommerfeld's thirteen doctoral students during this period, only Alfred Landé dealt with quantum problems (Table A.2).

Yet the quantum was never far away whenever a problem withstood classical approaches. Sommerfeld was among the first who responded to Bohr's quantum

---

[1] For a list of Sommerfeld's publications, see the bibliography in ASGS IV.

model of the atom. Bohr sent Sommerfeld a reprint of his first paper, which appeared in the *Philosophical Magazine* in July 1913, because he expected that his theory would be met enthusiastically in Munich (interest elsewhere had been rather meager). "I thank you very much for sending me your highly interesting work which I had already studied in the *Phil. Mag.*," Sommerfeld replied in September 1913. "The problem of expressing the Rydberg-Ritz constant by Planck's h has been on my mind for a long time. I discussed it with Debye a few years ago. Although I am skeptical about atomic models in general, the calculation of this constant is without doubt a great accomplishment." He also mentioned another reason why Bohr's model attracted his interest at that moment, "Will you also apply your atomic model to the Zeeman effect? I would like to research the topic in more depth."[2]

Sommerfeld's interest in the Zeeman effect, the splitting of spectral lines in a magnetic field, was kindled by a discovery made by Friedrich Paschen and his doctoral student Ernst Back in 1912. They showed that the pattern of line-splitting transforms into a different pattern at very high magnetic field strengths. Sommerfeld attempted to explain this transformation by extending Lorentz's concept. Lorentz won the Nobel Prize with Zeeman in 1902 for his classical theory on the Zeeman effect. Sommerfeld replaced Lorentz's assumption of isotropic electronic oscillations with anisotropic oscillations—allowing an electron to vibrate at different frequencies in different directions (Sommerfeld 1913). Sommerfeld's theory provoked a dispute with Voigt, the Göttingen theorist, who regarded himself as the architect of the theory of anisotropically bound electrons. In a sequel, Sommerfeld generalized Voigt's approach (Sommerfeld 1914). But the theory involved rather arbitrary assumptions. Bohr explained the emission of spectral lines in an entirely different manner. However, with regard to the splitting of spectral lines in magnetic fields, resorting to Bohr's model offered no alternative—at least not in its present state.

Another riddle for which Bohr's model seemed applicable was the splitting of spectral lines in electric fields, called the Stark effect (Leone et al. 2004). "Stark was recently here," Sommerfeld reported enthusiastically to Wien in November 1913, "and presented a really great discovery to us. An electrical Zeeman effect of a quite large order."[3] On December 10, 1913, the Stark effect was the subject of further discussion in the Munich Wednesday Colloquium. On May 13, 1914, two colloquia addressed the Stark effect from the perspectives of an experimentalist (Wagner from

---

[2]Sommerfeld to Bohr, 4 September 1913, NBA (Bohr). Also in ASWB I, doc. 202. "Ich danke Ihnen vielmals für die Übersendung Ihrer hochinteressanten Arbeit, die ich schon im Phil. Mag. studirt hatte. Das Problem, die Rydberg-Ritz'sche Constante durch das Planck'sche h auszudrücken, hat mir schon lange vorgeschwebt. Ich habe davon vor einigen Jahren zu Debye gesprochen. Wenn ich auch vorläufig noch etwas skeptisch bin gegenüber den Atommodellen überhaupt, so liegt in der Berechnung jener Constanten fraglos eine grosse Leistung vor [...] Werden Sie Ihr Atommodell auch auf den Zeeman-Effekt anwenden? Ich wollte mich damit beschäftigen." For a detailed account of Sommerfeld's extension of Bohr's model from 1913 to 1916, see (Eckert 2013b, 2014); an extensive study on the development of Bohr's atomic theory prior to quantum mechanics is given in (Kragh 2012).

[3]Sommerfeld to Wien, 29 November 1913, DM, NL 56, 010. "Kürzlich war J. Stark hier und hat über eine ganz grosse Entdeckung vorgetragen. Elektrischer Zeeman-Effekt ganz grosser Ordnung [...]."

Röntgen's institute) and a theorist (Lenz from Sommerfeld's institute). Two weeks later, Sommerfeld and Lenz presented Bohr's recently published approach on the Stark effect in the Wednesday Colloquium. At the end of the summer semester, Bohr visited Munich and lectured at the Wednesday Colloquium on "the spectra of helium and hydrogen." In the following winter semester, Sommerfeld dedicated his special lecture to the subject of the Zeeman effect and spectral lines. On January 16, 1915, Sommerfeld again reported on the Stark effects in the colloquium. It was quite obvious that the Zeeman and the Stark effects became Sommerfeld's proving ground for Bohr's views.[4]

In the meantime, the First World War caused some changes to the routine operation of Sommerfeld's institute. Students and younger faculty were drafted for war service. Sommerfeld had to run his institute without his assistants, Friedrich and Lenz. At the same time, he became involved in war research, a digression which he used as an opportunity to pursue some of his more mathematically-oriented work, such as the spread of electromagnetic waves. Yet Sommerfeld's institute preserved its character as a nursery. His students wrote moving letters from their various sites of deployment, expressing a growing desire to stay in contact with recent advances in physics as the long war dragged on. Hopf, for example, felt "a ravenous appetite for physics" when he learned about Sommerfeld's excitement about Einstein's recent advances in the theory of gravitation.[5]

Bohr's theory, therefore, lost none of its allure—in Munich or in the trenches. Enthusiasm for atomic theory mixed with anxiety about the war. "I lectured this semester on Bohr and I am extremely interested in it, as far as the war permits," Sommerfeld wrote to Wien on 22 February 1915. On the same day, the newspapers reported a German victory against the Russian army in East Prussia. "The news of the 100,000 Russians today is indeed even more wonderful than Bohr's explanation of the Balmer series," Sommerfeld rejoiced. In the same paragraph, he mentions that he is extending Bohr's theory for the first time. "I have wonderful new results on this matter."[6]

Whatever he may have accomplished, Sommerfeld did not rush to publish it. In March 1915, Ludwig Prandtl, director of the University of Göttingen Institute for Applied Mechanics, requested Sommerfeld's help with war research concerning submarines and the propagation of sound under water—problems for which Sommerfeld was prepared "better than any other person," as he responded to Prandtl.

---

[4]Physikalisches Mittwoch-Colloquium, DM, 1997–5115; list of Sommerfeld's lectures in Table A.4.

[5]Hopf to Sommerfeld, 13 November 1915, DM, NL 89, 059. "[…] hoffentlich lässt sich der Heisshunger nach Physik bald in einer besseren Zeit stillen; ich habe gar nichts dagegen, wenn ich den Frieden in meiner jetzigen Stellung erleben darf, grosse Lorbeeren kann ich mir als Krieger doch nicht holen. Einstein sprach ich in Berlin nur einmal; ich freue mich sehr über Ihre Begeisterung für die Gravitationstheorie; denn ich war von Anfang an davon überzeugt, dass dies die Krönung seines bisherigen Lebenswerks sein würde." For other war-related aspects, see ASWB I, 445–455.

[6]Sommerfeld to Wien, 22 February 1915, DM, NL 056. Also in ASWB I. "Ich habe in diesem Semester über Bohr gelesen und bin äusserst dafür interessirt, soweit der Krieg es zulässt. Die heutigen 100000 Russen sind freilich noch schöner wie die Erklärung der Balmer'schen Serie bei Bohr. Ich habe schöne neue Resultate dazu."

At the same time, he revealed that Wien's cousin, Max Wien, had also requested his help with problems of "military hydrodynamics."[7] But he must have pursued his investigations on atomic theory and shared the results by corresponding with his assistant Lenz, who was in the trenches in Northern France. "I very much enjoyed your discovery concerning Bohr's model and the Stark effect," Lenz answered in April 1915, "and I am curious about its further development."[8] In another letter a month later, Sommerfeld also revealed to Wilhelm Wien that "Bohr's theory of the hydrogen lines" led him to "an interesting approach for the Stark effect," a subject on which he had lectured in the last semester. But "problems of war physics" and other obligations prevented him from elaborating.[9] During the summer semester in 1915, he allowed himself another digression by dedicating his special lecture to Einstein's general theory of relativity, about which he was "as much excited as last semester about Bohr."[10] In the Wednesday Colloquium, Bohr's theory was also a recurring topic. Walther Kossel, an assistant from the physics institute of the Technical University in Munich, gave a special lecture on June 30, 1915 that focused on "chemical formulae from the perspective of using Bohr's model." According to Sommerfeld, at that time Kossel was developing a paper "concerning chemical valences from Bohr's view." Sommerfeld recommended this work to Wien in his capacity as the editor of the *Annalen der Physik*. Although Kossel was not Sommerfeld's doctoral student, he became a close participant in Sommerfeld's circle, focusing on X-ray spectroscopy as his specialty. "In the field of X-ray frequencies, Kossel has decisive merits, which are also acknowledged by Bohr," Sommerfeld praised of his colleague.[11]

By the end of 1915, Sommerfeld finally elaborated on ideas only previously alluded to in his correspondence. He felt prompted to do so by a recent paper of Bohr's, on which Sommerfeld had reported in the Munich colloquium on November 27, 1915. Around the same time, Paschen wrote to Sommerfeld that his measurement of the spectrum of ionized helium had led him to the conclusion "that Bohr's theory

[7]Prandtl to Sommerfeld, 11 March 1915, GOAR 2666. Sommerfeld to Prandtl, 14 March 1915, GOAR 2666. "Ich werde mich natürlich sehr gern und sogleich an Ihre Probleme machen. Ich glaube auch, dass ich sie bewältigen kann, jedenfalls leichter und vollständiger als ein Anderer. Ich habe auch von Berlin aus (durch Max Wien) ähnliche Fragen vorgelegt bekommen—ähnlich nur insofern, als sie sich auch auf die militärische Hydrodynamik beziehen."

[8]Lenz to Sommerfeld, 10 April 1915, DM, NL 89, 059. "Über Ihre Entdeckung zum Bohrmodell und Starkeffekt habe ich mich sehr gefreut und bin auf den weiteren Fortgang sehr gespannt."

[9]Sommerfeld to Wien, 3 May 1915, DM, NL 056. Also in ASWB I. "Ich habe im vorigen Semester einen interessanten Ansatz für den Stark-Effekt aus d. Bohr'schen Theorie der Wasserstofflinien gewonnen. Es fehlt aber noch an der Durchführung, weil mir teils Probleme der Kriegsphysik, teils ein Beitrag zur Elster-Geitel-Festschrift dazwischen gekommen sind."

[10]Sommerfeld to Schwarzschild, 31 Juli 1915, SUB, Schwarzschild 743. "Ich habe in diesem Semester Relativität, zuletzt im Sinne der Einstein'schen letzten Berliner Arbeit, gelesen und bin sehr davon begeistert, fast so wie im vorigen Semester von Bohr."

[11]Sommerfeld to Wien, 31 December 1915, DM, NL 56, 010. "Ich hätte Ihnen auch gern ein empfehlendes Wort für eine dicke Arbeit von Kossel gesagt, die er an die Annalen schicken wollte, betr. Chemische Valenzen vom Bohr'schen Gesichtspunkt. [...] Auf dem Gebiet der X-Strahl-Frequenzen hat Kossel ja entschiedene auch von Bohr anerkannte Verdienste." On Kossel's contributions, see (Heilbron 1967).

is exactly correct, apart from the complicated structure of the lines 4686 etc."[12] This information must have pushed Sommerfeld to publish what he had achieved so far, because he was confident that his extension of Bohr's model explained not only the line splitting in the Zeeman and Stark effects but also Paschen's "complicated structure of the lines 4686" in the helium spectrum. On December 6, 1915 and January 8, 1916, he presented this elaboration in a preliminary form to the Bavarian Academy of Science (Sommerfeld 1915a, b). For the comprehensive publication, he again took his time; he did not submit it to the *Annalen der Physik* until July 1916 (Sommerfeld 1916a).

It suffices to briefly summarize the main achievements at this stage.[13] Sommerfeld's approach was to quantize not only the azimuthal motion of an electron around the atomic nucleus, like in Bohr's model, but also its radial motion. This assumption implied that the electron's motion had to be treated relativistically because it was allowed to accelerate to high velocities when it approached the nucleus. Without relativity, the azimuthal and radial motions resulted in a stationary Kepler orbit; with relativity, the orbit became a precessing ellipse. Quantizing the non-relativistic motion with two quantum conditions for the radial and azimuthal action integrals yielded Bohr's result for the energy levels of the orbiting electron, except that there was now a total of two quantum numbers whereas Bohr had only one. In the non-relativistic case, the two quantum numbers appeared only as a sum so that they could be replaced by a single quantum number. This did not only recover Bohr's formulae, but also hinted at the degeneracy of the problem. In the relativistic case, the quantum numbers could no longer be replaced by a single one; specifically, the energy levels of the orbiting electron, which fell together in the non-relativistic case, became split in different levels. This gave rise to the "fine structure" to which Paschen alluded when he mentioned "the complicated structure of the lines 4686." Furthermore, it offered an explanation for the doublet structure in the spectra of hydrogen and hydrogen-like heavier atoms. The relativistic effect was proportional to the fourth power of the atomic number Z so that what was hardly observable in light elements became magnified for heavier elements. "The application that our concept finds in the K- and L-series of X-rays is particularly surprising," Sommerfeld hinted at the new field of X-ray spectroscopy as an unexpected proving ground for his theory. The so-called L-doublets, which had been observed across the entire periodic system, could be considered a magnified image of the hydrogen doublets, with the magnification factor proportional to the fourth power of the atomic number. For a heavy atom like gold, the magnification was 37,000,000-fold. In contrast to the hydrogen doublets, which could be resolved only using sophisticated experimental techniques, the X-ray doublets were "distinct and widely separated lines that could be measured in this range of frequencies with entirely sufficient preci-

---

[12]Paschen to Sommerfeld, 24 November 1915, DM, HS 1977–28/A,253. Also in ASWB I. "Schon jetzt sehe ich, dass Bohr's Theorie exact richtig ist abgesehen von der complicirten Structur der Linien 4686 etc."

[13]The extension of Bohr's model has long been a focus of historians from various perspectives (Jammer 1966, Chap. 3; Nisio 1973; Benz 1975; Kragh 1985; Robotti 1986). For a documentary history and further references, see ASWB I, 431–445 and doc. [197]–[265].

sion, despite the still rather primitive means of observation" (Sommerfeld 1915b, 460).[14] Henry Moseley's preliminary comparison with X-ray spectra reflects a good balance between theory and experiment. In short, Sommerfeld's theory not only provided support for the basic assumptions underlying Bohr's model, it also extended its scope to new horizons. "Your spectral investigations are among the most beautiful experiences I have had in physics. Only through your work is Bohr's idea fully convincing," Einstein congratulated.[15]

The published articles suggest that the Bohr-Sommerfeld model was a seed planted by Bohr and brought to fruition by Sommerfeld in a combined effort. In the absence of advanced students and assistants and with Sommerfeld involved in war research himself, atomic theory did not seem to reflect the involvement of Sommerfeld's nursery—at least for the duration of the war. However, a closer inspection reveals that this was not the case. Sommerfeld's extension of Bohr's model was not a single feat that culminated in the publication of the seminal *Annalen* paper (Sommerfeld 1916a). Within three years—between its inception in 1914–1915 to the end of the war—the theory went through several modifications. Furthermore, both before and after the *Annalen* publication in 1916, Sommerfeld's theory was a group effort to some extent, involving both experimental and theoretical contributions. Before the war, networking had already become an integral element for the operation of Sommerfeld's school. What is labeled as the Bohr-Sommerfeld model of the atom, therefore, is not a single achievement, but rather a complex process that extends over a couple of years and involves more than these two main actors.

Without going into technical detail, it seems appropriate to illustrate this group effort with a few examples. One contributor was Lenz; Sommerfeld sent his first Bavarian Academy memoirs (Sommerfeld 1915a, b) to Lenz on war deployment in Northern France. Lenz simplified Sommerfeld's somewhat tedious expressions and derived the law that corresponds to the Balmer formula in the relativistic case: "Because this law was not explicitly presented in your work," Lenz wrote to Sommerfeld in March 1916, "I derived it myself in the following manner [...]"[16] Sommerfeld acknowledged his assistant's contribution from the trenches when he declared that he was alerted to "the closed form of the spectral equation in an army

---

[14]"Besonders überraschend ist die Anwendung, welche unsere Auffassung im Gebiete der K- und L-Serie der Röntgenstrahlung findet. Hier treten durch das ganze natürliche System der Elemente hindurch von $Z = 34$ bis $Z_1 = 80$ (Z = Ordnungszahl des Elementes = Stellenzahl im natürlichen System) Dubletts auf, die denselben Ursprung haben wie die Wasserstoffdubletts, und geradezu als ein um den Betrag $(Z - l)^4$ vergrößertes Abbild jener anzusehen sind. Der Größe dieses Faktors $(37 \cdot 10^6$ bei Gold) ist es zu verdanken, daß namentlich in der L-Serie diese Dubletts nicht mehr unter die unscheinbaren Merkmale der Feinstruktur fallen, sondern als verschiedene, weit getrennte Linien beschrieben wurden und trotz der vorläufig naturgemäß noch primitiven Beobachtungsmittel in diesem Frequenzbereich mit völlig ausreichender Genauigkeit gemessen werden konnten."

[15]Einstein to Sommerfeld, 3 August 1916, DM, HS 1977–28/A,78. Also in ASWB I. "Ihre Spektral-Untersuchungen gehören zu meinen schönsten physikalischen Erlebnissen. Durch sie wird Bohrs Idee erst vollends überzeugend."

[16]Lenz to Sommerfeld, 7 March 1916, DM, NL 89, 059. Also in ASWB I.

postal letter from W. Lenz" in his *Annalen* paper (Sommerfeld 1916a, 54).[17] The fact that Sommerfeld mentioned it in correspondence with other colleagues shows how much he valued Lenz's contribution.[18]

Another early contributor was Karl Glitscher, who succeeded Friedrich as Sommerfeld's assistant in 1916. Glitscher made Sommerfeld's theory the subject of his doctoral thesis. He analyzed the fine-structure splitting by comparing Sommerfeld's relativistic calculation with an alternative approach based on the older electron theory (which also involved a dependency of the electron mass from the velocity). He arrived at the result that only the former yields agreement with experiments. The pre-relativistic electron theory, therefore, "is to be definitively rejected," Sommerfeld praised Glitscher's achievement in his doctoral report to the faculty.[19] Thus the fine-structure theory turned into a confirmation of Einstein's theory of relativity.

While these contributions merely added one theoretical facet or another, others extended the scope of the theory and gave rise to unforeseen ramifications in theoretical and experimental areas. In his presentation before the Bavarian Academy of Science, Sommerfeld acknowledged that "Mr. Kossel's colloquium talk has led me to apply my theory to X-ray frequencies" (Sommerfeld 1915a, 492).[20] At this opportunity, Kossel also informed the Munich physicists to a recent dissertation by a disciple of Manne Siegbahn whose laboratory focused on X-ray spectroscopy. Henceforth, Kossel became Sommerfeld's local addressee and collaborator in all questions related to X-ray spectra (Heilbron 1967). At the same time, Siegbahn and his experimental school in Lund, Sweden entered Sommerfeld's network as a foreign correspondent (Kaiserfeld 1993). Siegbahn and his doctoral students played a similar role for Sommerfeld in the rising field of X-ray spectroscopy, as Paschen's laboratory in Tübingen did for spectroscopy at optical frequencies.

The interaction of the Munich group with external contributors was also crucial for applying the theory to the Zeeman and Stark effects. Sommerfeld wished to apply Bohr's theory to these two effects shortly after he learned of it in the summer of 1913.

---

[17] "Auf die vorstehende geschlossene Form der Spektralgleichung bin ich durch einen Feldpostbrief von W. Lenz aufmerksam gemacht worden. In meiner ursprünglichen Darstellung hatte ich die im nächsten Paragraph vorzunehmende Potenzentwicklung nach a schon an einer etwas früheren Stelle eintreten lassen, wobei die Übersichtlichkeit und Geschlossenheit der Spektralformel verloren ging."

[18] Sommerfeld to Runge, 6 September 1916, DM, HS 1976-31. Also in ASWB I. "[…] und die geschlossene Spektralformel für die Balmerserie steckt implicite in meiner ersten Arbeit; ich habe dort nur etwas zu früh nach Potenzen entwickelt; in meiner Annalenarbeit, die in diesen Tagen erscheinen muß, kommt sie explicite vor, unter Citirung eines Feldpostbriefes meines Assistenten W. Lenz."

[19] Sommerfeld, Report to the faculty on Glitscher's dissertation, 15 February 1917. UAM, OC I 43 p. "[…] definitiv zu verwerfen."

[20] Kossel's contribution concerned a relationship between the X-ray doublets of the L- and K-series. "Diese Feststellung Kossels ist unabhängig von meiner Theorie erfolgt und hat mich umgekehrt, bei Gelegenheit eines Colloquium-Vortrages von Hrn. Kossel, dazu geführt, meine Theorie auf die Röntgen-Frequenzen anzuwenden." Kossel's remark was presumably part of his colloquium talk on ?? December 1915 on the "Dissertation von Malmer betr. Röntgenspektren." Physikalisches Mittwoch-Colloquium, DM, 1997-3115.

In his attempt to explain the Paschen-Back effect classically, he had argued that the anomalous and the normal Zeeman effects are due to a degeneracy. Lorentz's theory about the normal Zeeman effect assumes that a spectral line is caused by three equal quasi-elastic vibrational modes for an electron, which could oscillate along three vertical axes. In a magnetic field, the line was split because the degeneracy was resolved. He recalled this effort when he outlined his ideas on Bohr's model before the Bavarian Academy: "The magnetic field does not create new vibrational modes, it only decomposes those that already exist. The originally coinciding frequencies appear as a more unstable structure, which is easier to influence than the distinct frequencies of an originally anisotropic vibrating electron." What was true for a quasi-elastically vibrating electron should also apply to an orbiting electron. If a spectral line in Bohr's model was a coinciding ensemble of lines that could be decomposed into a fine-structure by some cause that resolves the degeneracy, the line splitting of the Zeeman and Stark effects could be explained in the same manner as the doublets and more complicated fine structures. "This view is confirmed by the increasing number of components, which grows with the number of the Balmer line," he concluded with a reference to a recent Stark publication (Sommerfeld 1915a, 449).[21] All attempts to explain the Stark effect pattern, including a recent effort by Bohr, had failed in just this regard, specifically, they could not account for the growing number of components. However, the elaboration of a theory for the Stark and Zeeman effects met with "the difficulty to apply the quantum approach to non-periodic orbits into which the Keplerian ellipses become distorted" (Sommerfeld 1915a, 426).[22] Therefore, Sommerfeld could not yet offer a solution for what he originally had in mind as his major goal.

This deficiency was overcome in relation to the Stark effect by Epstein and Schwarzschild in the course of a fierce rivalry in the spring of 1916. When the war broke out, Epstein had just completed his doctoral work with Sommerfeld. He intended to move to Zurich where he was offered the possibility of becoming a Privatdozent, but the war thwarted this plan. "Germany was practically sealed off; and I was an enemy alien and not permitted to leave Germany. I had to stay in

---

[21]"Die elementare Lorentzsche Theorie des Zeeman-Effektes beruht darauf, daß in jeder Spektrallinie drei unter sich gleiche Hauptschwingungen eines quasielastisch-isotrop schwingenden Elektrons übereinanderfallen. Das Magnetfeld erzeugt keine neuen Schwingungsmöglichkeiten, sondern legt nur die vorhandenen auseinander. Die ursprünglich zusammenfallenden Frequenzen erscheinen dabei als ein labileres, durch äußere Einwirkung leichter zu beeinflussendes Gebilde wie die ursprünglich verschiedenen Frequenzen eines anisotrop schwingenden Elektrons [...] Diese Anschauung überträgt sich unmittelbar auf den Stark-Effekt bei der Balmer-Serie [...] Für diese Auffassung des Stark-Effektes spricht die große und mit der Nummer der Balmer-Linie steigende Komponentenzahl [...]" Sommerfeld had reported in the Munich colloquium about Stark's recent findings on 16 January 1915 "Die Anzahl Zerlegungen beim Starkeffekt des Wasserstoffs, Stark, Berl. Akad." Physikalisches Mittwoch-Colloquium, DM, 1997–5115.
[22]"Diese Absicht scheiterte indessen vorläufig an der inzwischen auch von Bohr stark betonten Schwierigkeit, den Quantenansatz anzuwenden auf nicht-periodische Bahnen, in welche ja die Keplerschen Ellipsen durch ein elektrisches Feld auseinander gezogen werden."

Munich the whole time," he recalled later.[23] Although he was formally a prisoner, he was allowed to work in Sommerfeld's institute. "I needed what is called a *Habilitationsschrift*—a thesis for becoming a Privatdozent. And that was after Sommerfeld had published that fine structure paper. So I said to Sommerfeld that I would take the Stark effect," Epstein reported about this phase of his early career. Then Sommerfeld informed him that he had also encouraged Schwarzschild to take up the same problem. "Now I was a little crestfallen, because I regarded this as a stab in the back," Epstein told his interviewer. But he had enough familiarity with the problem to meet the challenge:

> You see, I knew already how the electron moves, and I knew how to do it. I got up at 5 o'clock the next morning and by 10 I had the formula. And then the same morning I brought it to Sommerfeld. And what do you know, the same afternoon he got a letter from Schwarzschild, and Schwarzschild had the wrong formula. It was the same order of magnitude, but didn't agree on the positions of the lines. Sommerfeld wrote Schwarzschild, "This morning Epstein brought me the formula of the Stark effect, and this afternoon we got your letter. But Epstein's formula agrees with the observation." When Schwarzschild got [his result], he immediately announced in the Berlin Academy that he would speak about it. And he did, before he wrote the letter to us, so he reported it wrong in the Academy. By that time however I had already sent my announcement that came out just one day before he delivered that lecture in the Academy. But in the lecture he corrected it, and in the proof he removed all the discrepancies, and it came out correct also. Of course the two final papers came out much later. So I had the priority by one day.[24]

Although Epstein's recollection seems to dramatize his rivalry with Schwarzschild, it is by and large confirmed by the archival record. Both used methods from celestial mechanics. "I am very impressed that you cavort in Belgium and in quantum heaven at the same time," Sommerfeld wrote in a response to a letter in which Schwarzschild, who was drafted for war service in Belgium, had reported how he approached the problem by using action-angle-variables. "Although I am not familiar with your notions from general celestial mechanics [...] I think that our views are not far apart from another."[25] Schwarzschild reformulated Sommerfeld's quantization scheme by adapting the Hamilton-Jacobi formalism to quantum problems with the more general action integrals instead of Sommerfeld's original phase integrals as the appropriate quantities for quantization. In a four-page letter dated March 21, 1916, he reported how he had solved the Stark effect problem.[26] Sommerfeld responded three days later that Epstein had, at the same time, arrived at a "more general formula" that also

---

[23]Interview with Epstein by Alice Epstein, 22 November 1965. Oral History Project, California Institute of Technology Archives, Pasadena, California, http://oralhistories.library.caltech.edu/73/, accessed 17 April 2020.

[24]Interview with Epstein by Heilbron, 25 May 1962. AHQP, https://www.aip.org/history-programs/niels-bohr-library/oral-histories/4592-1, accessed 17 April 2020.

[25]Sommerfeld to Schwarzschild, 9 March 1916, SUB, Schwarzschild 743. Also in ASWB I. "Dass Sie sich gleichzeitig in Belgien und im Quantenhimmel tummeln, imponirt mir sehr. Wenn mir auch Ihre Begriffe aus der allgemeinen Himmelsmechanik (die eindeutigen Winkelcoordinaten w_k) nicht geläufig sind, so glaube ich doch, dass unsere Auffaßungen nicht weit auseinander gehen."

[26]Schwarzschild to Sommerfeld, 21 March 1916, DM, HS 1977–28/A,318. Also in ASWB I.

contained a line missing in Schwarzschild's result.[27] Epstein's and Schwarzschild's preliminary communications were submitted for publication a week later (with one day difference). Schwarzschild died a few weeks later from a skin disease. Sommerfeld commemorated him in a paper titled "The Quantum Theory of Spectral Lines and the Last Work of Karl Schwarzschild" (Sommerfeld 1916b).

With Epstein's and Schwarzschild's work, the somewhat haphazard schemes of quantizing used by Bohr, Sommerfeld, and more recently, Planck (Eckert 2008) assumed a more coherent form that could be used to solve quantum problems. First, the equations of motion of the mechanical problem needed to be translated into the language of Hamilton-Jacobi formalism. The first (and most difficult) part of this procedure was to find out in which coordinates the classical problem was separable, specifically, to find the canonical form of the equations of motion. The "quantum" part of the problem was then fairly straightforward following Hamilton-Jacobi formalism. Sommerfeld first applied this procedure to the problem of the Zeeman effect. He had already indicated in his *Annalen* paper (Sommerfeld 1916a) that the plane orbit of an electron becomes spatially deformed by an applied magnetic field. The Hamilton-Jacobi approach to this problem led to phase integrals that were subject to complex integration—Sommerfeld's specialty since his period in Göttingen as a mathematician. In September 1916, Sommerfeld submitted a paper to the *Physikalische Zeitschrift* (chosen for rapid publication to ensure high priority) where he presented this procedure for the Zeeman and the Stark effects, comparing both non-relativistic and relativistic motion. He also used the opportunity to simplify Epstein's and Schwarzschild's treatments of the Stark effect. With its focus on Hamilton-Jacobi formalism and complex integration, the paper presented concrete examples so that it could be taken as a role model for further applications (Sommerfeld 1916c).

Debye, who had in the meantime moved to Göttingen as director of the Physical Institute, published practically the same results on the Zeeman effect in the same issue of the *Physikalische Zeitschrift*. They were obtained by basically the same procedure and further emphasized this method of quantization (Debye 1916). As Debye later recalled, he felt that Sommerfeld "did not like" that he also worked in this field. "He wanted to have that for himself. So I decided, well all right, I won't do that anymore." Feelings of rivalry and reverence must have been in frequent conflict among Sommerfeld's pupils. Debye gave preference to the latter feeling and left this territory to Sommerfeld: "I really owed him so much."[28]

With regard to the physical meaning, Sommerfeld's and Debye's treatments of the Zeeman effect involved introducing a spatial quantization by a third quantum number. For hydrogen, it resulted in a splitting of each line into a triplet just like Lorentz's classical theory (quantitatively, the splitting was not exactly the same). The relativistic treatment did not change this result. If relativity resulted in a fine-structure split without a magnetic field, why not with a magnetic field? Although the

---

[27] Sommerfeld to Schwarzschild, 24 March 1916, SUB, Schwarzschild 743. Also in ASWB I.

[28] Interview with Debye by Kuhn and Uhlenbeck, 3 May 1962, https://www.aip.org/history-programs/niels-bohr-library/oral-histories/4568-1, accessed 17 April 2020.

experimental results at the time did not provide enough evidence to determine whether hydrogen showed a normal or anomalous Zeeman effect, Sommerfeld expressed some suspicion about his own results. A transformation of the pattern as observed in the Paschen-Back effect seemed beyond reach if each line was affected in the same manner (Kragh 1985; Jensen 1984; Robotti 1992). This conclusion seemed to indicate a serious problem. When Ehrenfest congratulated Sommerfeld and Epstein and wished "Munich physics" success in the pursuit of this research program,[29] Sommerfeld expressed some confidence about the scheme of quantization, "which is actually that of Epstein, but also agrees with Planck." The only qualification he made concerned the Zeeman effect, in which "uncertainty prevails. In contrast to Debye, I consider the result of the quantum theory concerning the Zeeman effect to be wrong. If you could clarify this by means of your adiabatic hypothesis, we would be very grateful."[30]

This exchange illustrates that by the fall of 1916 the "old quantum theory" (as it is often summarily designated) entered a stage of corroboration, scrutiny, and further extension. Ehrenfest and his doctoral student, Johannes Martinus Burgers, payed tribute to this development by focusing on adiabatic invariants—quantities that remained constant when other quantities of a system underwent slow changes. By the end of the year, it was clear that the quantities which Bohr, Sommerfeld, Epstein, Planck, and Schwarzschild had discerned in one way or another for the quantization of a mechanical system were adiabatic invariants. In three sessions of the Amsterdam Academy in December 1916 and January 1917, Burgers presented a trilogy of papers detailing the connection between the adiabatic invariants of mechanical systems and their quantization (Péres 2009). On March 2, 1917, Epstein reported on these results at the Munich colloquium.[31] At the same time, Sommerfeld applied the quantization of adiabatic invariants to the motion of electrons in molecules in an attempt to account for the dispersion of light. In contrast to atoms, where electrons do not radiate in their stationary orbits, Sommerfeld argued that the orbiting electrons in molecules are slightly deflected from their stationary orbit by light. If the oscillation of light was slow compared to the speed of the electrons in their orbits, the changes in the orbital motion could be treated as adiabatically reversible, Sommerfeld argued. "Hence, if

---

[29]Ehrenfest to Sommerfeld, undated [April/May 1916], DM, HS 1977–28/A,76. Also in ASWB I. "Begreiflicherweise hat meinen Freunden und mir Ihre Arbeit und der daran anschließende Erfolg von Epstein sehr große Freude bereitet. So entsetzlich ich es auch finde, dass dieser Erfolg //vorläufig// wieder dem vorläufig doch noch so ganz kanibalischem Bohr-Modell zu neuen Triumphen verhilft "dennoch wünsche ich der Münchner Physik herzlich weitere Erfolge auf diesem Weg!".

[30]Sommerfeld to Ehrenfest, 16 November 1916, MBL. Also in ASWB I. "Ich fühle mich auf dem in den Ann. eingenommenen Standpunkt der Quantelung, welcher eigentlich der Epstein'sche ist, aber auch mit Planck durchweg übereinstimmt, sicher, zumal ja auch die Unsicherheit in der Coordinatenwahl beseitigt ist. Nur im Punkte des Zeeman-Effektes herrscht Unsicherheit. Ich sehe nämlich, im Gegensatz zu Debye, das Ergebnis der Quantentheorie des Zeeman-Effektes als falsch an. Wenn Sie hier mit der Adiabaten-Hyp. Klarheit schaffen könnten, wäre das sehr dankenswert."

[31]Physikalisches Mittwoch-Colloquium, DM, 1997-5115.

we calculate the orbital changes by light mechanically into the theory of dispersion, we are not contradicting quantum theory" (Sommerfeld 1917, 502).[32]

Sommerfeld's dispersion theory met with severe criticism and was soon abandoned, but it illustrates how the new concepts developed in 1916, such as the quantization of adiabatic invariants, were adopted by Sommerfeld and his circle as revolutionary instruments for the exploration of the new quantum territory. Optical dispersion, however, remained a problem for which there was no answer (Duncan and Janssen 2007; Jordi 2018; Jähnert 2019). Dispersion was not the only frustrated effort. After his success with the Stark effect, Epstein attempted to quantize hyperbolic orbits. If stationary hyperbolic orbits existed, an electron that happened to be put in such an orbit, for example by absorbing light or by radioactive decay, would be ejected from the orbit of the atom as a photoelectron or beta ray (Epstein 1916).

The entries in the Munich colloquium book document an increasing interest in quantum themes.[33] Epstein presented his quantization of hyperbolic orbits together with his theory of the Stark effect on two occasions. Sommerfeld reviewed the recent state of his own theory, including the applications to X-ray spectra and the Zeeman effect—obviously a sort of rehearsal for a report that he presented to the Munich Academy on November 4, 1916 (Sommerfeld 1916d). At the same meeting, he also discussed Karl Ferdinand Herzfeld's statistical analysis of Bohr's model (Herzfeld 1916). In 1917, Sommerfeld twice reported on his dispersion theory. Furthermore, he introduced Einstein's "general formulation of quantum theory" to the Munich circle (Einstein 1917) in an attempt to explore the limits of the quantization schemes that served as the base for what became known later as EBK quantization, after Einstein, Brillouin and Joseph Keller (Bergia and Navarro 2000; Stone 2005).

Another modification of the "Bohr-Sommerfeld" theory was introduced when Bohr published a comprehensive treatise on "The Quantum Theory of Line Spectra." The work expanded upon a preliminary paper he had submitted for publication in 1916 but had withdrawn when he learned about the advances made in Munich. As he explained to Sommerfeld in an eight-page letter in 1919, he felt forced to modify his views continuously until he dared to submit this re-evaluation of the theory for publication. The main result of this effort was Bohr's correspondence principle, or *Analogieprinzip*, as he described it to Sommerfeld.[34] In the introduction to this memoir, written in November 1917, Bohr explained that he wished "to trace the analogy between the quantum theory and the ordinary theory of radiation as closely as possible."[35] The first tangible achievement of this principle was a set of selection rules, that is, a prescription to decide which transition (between the many stationary orbits that existed in the Bohr-Sommerfeld atom) were actually allowed to occur. The

---

[32]"Wenn wir also in der Dispersionstheorie die Bahnänderungen durch die Lichtwelle mechanisch berechnen, setzen wir uns nicht in Widerspruch mit der Quantentheorie."

[33]Physikalisches Mittwoch-Colloquium, DM, 1997–5115.

[34]Bohr to Sommerfeld, 19 November 1919, NBA, Bohr. Also in ASWB II. Part I and II of Bohr's memoir appeared in 1918, part III in 1922 (Nielsen 1976).

[35]The English translation is available at http://web.ihep.su/dbserv/compas/src/bohr18/eng.pdf, accessed 17 April 2020.

second achievement of the correspondence principle was a procedure for calculating the intensity of spectral lines (Darrigol 1992, 123–127; Jähnert 2019). Sommerfeld's theory so far only offered some plausibility arguments for both. Bohr's memoir was therefore "studied with enthusiasm," as Sommerfeld wrote to Bohr in May 1918. "Epstein reported about your work in the colloquium."[36]

Epstein's talk focused on just those two aspects that appeared to be the weak points of Sommerfeld's theory and where Bohr's approach appeared promising: the intensity and selection rules of spectral lines. The latter was also the subject of an investigation by Adalbert Rubinowicz, who had joined the Munich group as Sommerfeld's temporary assistant after being displaced from his home in Czernowitz.[37] Sommerfeld asked Rubinowicz to consider the problem of dispersion, which he had hoped in vain to solve for molecules by means of the adiabatic hypothesis. "In order to solve the problem posed by Sommerfeld," Rubinowicz later recalled, "I thought one should know more about the interaction of light and atoms."[38] He approached the problem using conservation laws that would have to be obeyed if an atom acts as an antenna for outgoing radiation and arrived at the same selection rules that Bohr had obtained from the correspondence principle. Rubinowicz therefore reacted "with mixed joy" to the news from Copenhagen, Sommerfeld confided to Bohr, because the explanation of the selection rules seemed to turn into a rivalry. Sommerfeld emphasized that he had already reported publicly about Rubinowicz's result on the occasion of Planck's sixtieth birthday in April 1918 in Berlin, so that the independence of both approaches was beyond dispute. Rubinowicz regarded the atom and the "ether" in which it was embedded as a combined system for which the conservation laws must hold when the atom emits radiation. "The atom provides only a certain quantum of energy and angular momentum," Sommerfeld summarized of Rubinowicz's approach. "The wave process happens in the ether only, which obeys Maxwell's equations […] By comparison of energy and angular momentum,

---

[36] Sommerfeld to Bohr, 18 May 1918, NBA (Bohr). Also in ASWB I. "Ihre Arbeit ist schon lange mit Spannung erwartet und wurde sofort von allen Seiten mit Eifer studirt. Da gerade Herr Flamm aus Wien hier zu Besuch war, trug uns Epstein über Ihre Arbeit im Colloquium vor." Epstein's colloquium talk on 11 May 1918 was about "Intensitätsfragen und Auswahlprinzip nach Bohr." Physikalisches Mittwoch-Colloquium, DM, 1997–5115.

[37] The University of Czernowitz (then belonging to the Austrian Empire, now called the Yuriy Fedkovych Chernivtsi National University) was closed when the war broke out Rubinowicz, who had been the assistant to the physics institute, spent his leave of absence (Beurlaubung) in Munich. See Interview with Adalbert Rubinowicz by Théo Kahan and Heilbron, 18 May 1963. AHQP.

[38] "Um das mir von Sommerfeld gestellte Problem zu lösen, dachte ich mir, müsste man zunächst etwas mehr über die Wechselwirkung zwischen Licht und Atom wissen. Es fiel mir ein, dass die Frequenzbedingung ein Erhaltungssatz war. Ich fragte mich daher, welche Erhaltungssätze man noch verwenden könnte. Als einziger kam da der Impulsmomentsatz in Frage. Ich habe daher auf Grund einer Arbeit von Abraham das ausgestrahlte Impulsmoment berechnet. Das war ungefähr im Oktober oder Sommer 1917. Kuhn: Die Aufgabe wurde Ihnen gestellt? Rubinowicz: Nein, nein, Sommerfeld hat mir keine Anweisung gegeben, wie ich das Problem anpacken soll. Er wollte nur, dass ich das Dispersionsproblem löse und bevor man etwa über die Dispersion aussagen kann, muss man doch etwas über die Ausstrahlung wissen und da es bei der Ausstrahlung einen Erhaltungssatz gibt, konnte man es mit einem anderen auch versuchen," See Interview with Adalbert Rubinowicz by Théo Kahan and Heilbron, 18 May 1963. AHQP.

Rubinowicz obtains a condition for the azimuthal quantum number."[39] Bohr's approach using the correspondence principle appeared to Sommerfeld "very efficient, although one does not really feel enlightened," as he wrote a few weeks later to Einstein.

> Some final remarks in Bohr's paper fit with a work by Rubinowicz, which has been submitted to the *Physikalische Zeitschrift* and which I mentioned to you recently. In my paper for my Planck speech, I stressed the matter somewhat further than in my oral presentation. Something like this: the atom does not vibrate, but rather the ether, whose business it is to vibrate. It does so in an entirely Maxwellian way, as it must do according to the amount of energy and angular momentum supplied by the atom.[40]

Rubinowicz, for his part, explained the simultaneity with and independence of his own method with Bohr's recent work in a footnote to his paper:

> When the present work was already complete, the first part of Bohr's treatise appeared (On the Quantum Theory of Line Spectra, Copenhagen Academy, 1918). In pursuit of this treatise, the same results will likely be obtained, based on the postulation that in the limit of long waves, the classical electron theory and the newer quantum theory must agree (Rubinowicz 1918).[41]

Along with the conceptual development, the institutional network was also expanding. Since the publication of Sommerfeld's first *Annalen* paper in 1916, the Munich nursery established and cultivated contacts in Berlin (Planck, Einstein), Tübingen (Paschen), Göttingen (Debye, Hilbert), Leiden (Ehrenfest), Copenhagen (Bohr) and Lund (Siegbahn). At each node of this network, one or another topic of the "Bohr-Sommerfeld" theory became part of a local research program. For the time

---

[39] Sommerfeld to Bohr, 18 May 1918, NBA (Bohr). Also in ASWB I. "Eine etwas gemischte Freude hatte über Ihre Arbeit Dr. Rubinowicz […]. Zur Feier von Plancks 60ten Geburtstag am 23. IV hatte ich in Berlin einen Vortrag gehalten, in dem ich auch die Frage streifte: Versöhnung von Quantentheorie und Wellentheorie […] Mein Standpunkt bei jenem Vortrage (bez. der von Rubinowicz in seiner Arbeit) ist Folgender: Der Wellenvorgang liegt allein im Äther, der den Maxwellschen Gleichungen gehorcht und quantentheoretisch wie ein linearer Oscillator wirkt, mit unbestimmter Eigenfrequenz ν. Das Atom liefert zu dem Wellenvorgang nur eine bestimmte Menge Energie und Impulsmoment als Material für den Wellenvorgang […] Durch Vergleich von Energie und Impulsmoment findet Rubinowicz eine Bedingung für die azimutale Quantenzahl: sie kann nur um höchstens eine Einheit sich ändern […]."

[40] Sommerfeld to Einstein, undated [June 1918], AEA (Einstein). Also in ASWB I. "Die neue Arbeit von Bohr haben Sie wohl gelesen. Seine Methode, Wellenth. u. Quantenth. von den grossen Quantenzahlen her aneinanderzupassen, scheint mir sehr wirkungsvoll, wenn sie einen auch innerlich nicht belehrt. Gewisse Schlussbemerkungen bei Bohr decken sich aber mit einer Arbeit von Rubinowicz, die inzwischen an die Phys. Ztschr. abgegangen ist u. von der ich Ihnen neulich schon sprach. Ich habe in dem Ms. zu meinem Planckvortrag die Sache etwas weiter ausgeführt, als ich es im Vortrag tun konnte. Etwa so: Nicht das Atom schwingt, sondern der Äther, dessen Metier es ist zu schwingen. Er tut dies ganz Maxwellisch, so wie er es nach dem vom Atom gelieferten Energie- und Impuls-Betrage tun muß."

[41] "Als die vorliegende Arbeit schon fertiggestellt war, erschien der erste Teil einer Abhandlung von Bohr (On the Quantum Theory of Line Spectra, Kopenhagener Akademie 1918), in deren Fortsetzung aus der Forderung, daß im Grenzfalle langer Wellen die klassische Elektronen- und die neuere Quantentheorie miteinander übereinstimmen müssen, u. a vermutlich auch die hier erhaltenen Ergebnisse abgeleitet werden dürften."

being, the interaction among the nodes was largely restricted to correspondence, but after the war, these relations gave rise to a vivid personal interaction in the form of visits and an exchange of traveling research fellows (see Sects. 6 and 7). The Munich quantum school—represented by Sommerfeld, Epstein, Glitscher, Rubinowicz, Kossel, and by correspondence, Sommerfeld's pupils drafted for war service—was a central node of this network. External honors such as a prize from the Prussian Academy of Science[42] and a call to Vienna (Eckert and Pricha 1984) further demonstrate Sommerfeld's centrality. "All of the honors that you have accumulated are well deserved," Hilbert congratulated Sommerfeld in February 1917.[43] Sommerfeld was compensated for declining the call to Vienna with a raise and the award of the title *Geheimrat* (privy councillor).[44]

At the same time, Sommerfeld made the new atomic research a subject of lectures addressed to scientists from all faculties. "I presented a one-hour popular lecture this semester on atomic structure and spectral lines," he wrote to Hilbert in March 1917, "which I also intend to publish as a book." Thus the seed was planted for *Atombau und Spektrallinien*, which appeared in 1919 and would soon be considered the bible of atomic physics.[45] He also presented a general lecture on atomic models to the Bavarian Polytechnical Association "in the presence of the king."[46] Early in 1918, the founder of the German Museum in Munich, Oskar von Miller, asked Sommerfeld to design an exhibit on the "newer theories" of atomic constitution of matter for his museum. Sommerfeld and his new colleague Kasimir Fajans, who was offered the position of physical chemist to Munich in 1917, eagerly accepted the request.[47] In the same year, the Royal Swedish Academy invited Sommerfeld to nominate a recipient for the 1918 Nobel Prize. Sommerfeld recommended "the

---

[42]Prussian Academy of Science to Sommerfeld, 25 January 1917, DM, NL 89, 020, Mappe 6,2. "[…] Ihnen die im laufenden Jahre zu vergebende Helmholtz-Prämie im Betrage von 1800 Mark zu verleihen als Anerkennung für die Arbeit 'Zur Quantentheorie der Spektrallinien'. Diese Verleihung ist in der heutigen zur Feier des Geburtsfestes Seiner Majestät des Kaisers und Königs und des Jahrestages König Friedrichs II gehaltenen Festsitzung öffentlich verkündet worden."

[43]Hilbert to Sommerfeld, 18 February 1917, DM, HS 1977–28/A,141. "[…] herzlichsten Glückwunsch für die Berufung nach Wien, über die ich mich wirklich ganz außerordentlich gefreut habe. Hoffentlich haben Sie für München auch ordentliche reelle Vorteile erreicht, die in jetziger Zeit besonders nötig sind. Ihre herrlichen Resultate und Theorien, insbesondere die Erklärung gewisser Linien im Röntgenspektrum durch Quantelungen hat uns hier Runge, und dann die Ihres Schülers Epstein hat uns hier Voigt referiert […] Ich vergass noch, Ihnen zur Helmholtzmedaille zu gratulieren—auf Sie häufen sich nun verdienter Massen alle Ehren!".

[44]Eugen von Knilling (Minister of Education in Bavaria) to Sommerfeld, 13 July 1917, DM, NL 89, 019, Mappe 5,5. Also in ASWB I. "[…] Titel und Rang eines K. Geheimen Hofrates […] daß mit Wirkung vom 1. Oktober lfd. Js. an der Ihnen zukommende Grundgehalt um 3000 M erhöht werde […]."

[45]Sommerfeld to Hilbert, 13 March 1917, SUB, Cod. Ms. Hilbert 379. "In diesem Semester habe ich ein einstündiges populäres Colleg über *Atombau und Spektrallinien* gelesen, vor etwa 80 Zuhörern, davon 12 Collegen, hauptsächlich Chemikern, Medizinern und Philosophen, das ich auch als Buch herausgeben möchte." For the history of this textbook, see (Eckert 2013c).

[46]Sommerfeld to Wien, 24 October 1917, DM, NL 56, 010. Also in ASWB I. "Letzten Montag musste ich über die Atommodelle im polytechn Verein in Gegenwart des Königs [vortragen]!".

[47]Miller to Sommerfeld, 28 January 1918; Sommerfeld to Miller, 31 January 1918, DM VA 1271.

quantum theory of Max Planck" for the Nobel Prize in 1918.[48] He may have thought that Bohr and he could not be awarded the prize if Planck's earlier merits were not first acknowledged in this manner. Whatever intentions and strategic goals he may have pursued, by 1918 Sommerfeld was in the center of the quantum network and its most ambitious promoter.

References

Benz U (1975) Arnold Sommerfeld: Lehrer und Forscher an der Schwelle zum Atomzeitalter, 1868–1951. Große Naturforscher 38. Wissenschaftliche Verlagsgesellschaft, Stuttgart

Bergia S, Navarro L (2000) On the Early History of Einstein's Quantization Rule of 1917. Archives Internationales d'Histoire des Sciences 50:321–373

Darrigol O (1992) From c-Numbers to q-Numbers: The Classical Analogy in the History of Quantum Theory. University of California Press, Berkeley. http://ark.cdlib.org/ark:/13030/ft4t1nb2gv/ (visited on 11/30/2020)

Debye P (1916) Quantentheorie und Zeemaneffekt. Physikalische Zeitschrift 17:507–6512

Duncan A, Janssen M (2007) On the Verge of Umdeutung in Minnesota: Van Vleck and the Correspondence Principle. Part 1. Arch Hist Exact Sci 61:553–624

Eckert M (2008) Plancks Later Work on Quantum Theory. In: Hoffmann D (ed) Max Planck: Annalen Papers. Wiley-VCH, Weinheim, pp 643–6652

Eckert M (2013b) Die Bohr-Sommerfeldsche Atomtheorie: Sommerfelds Erweiterung des Bohrschen Atommodells 1915/16. Klassische Texte der Wissenschaft. Springer, Berlin

Eckert M (2013c) Arnold Sommerfeld's Atombau und Spektrallinien. In: Badino M, Navarro J (eds) Research and Pedagogy: A History of Quantum Physics through Its Textbooks. Edition Open Access, Berlin, pp 117–135

Eckert M (2014) How Sommerfeld extended Bohr's model of the atom (1913–1916). Eur Phys J Hist 39:141–156

Eckert M, Pricha W et al (1984) Geheimrat Sommerfeld—Theoretischer Physiker. Deutsches Museum, Munich

Einstein A (1917) Zum Quantenansatz von Sommerfeld und Epstein. Verhandlungen der Deutschen Physikalischen Gesellschaft 19:82–92

Epstein PS (1916) Versuch einer Anwendung der Quantenlehre auf die Theorie des lichtelektrischen Effekts und der beta-Strahlung radioaktiver Substanzen. Annalen der Physik 50:815–840

Heilbron JL (1967) The Kossel-Sommerfeld Theory and the Ring Atom. Isis 58:450–485

Herzfeld KF (1916) Zur Statistik des Bohrschen Wasserstoffatommodells. Annalen der Physik 51:261–284

Jähnert M (2019) Practicing the Correspondence Principle in the Old Quantum Theory: A Transformation through Implementation. Springer, Berlin, Heidelberg

Jammer M (1966) The Conceptual Development of Quantum Mechanics. Mc Graw-Hill, New York

Jensen C (1984) The One-Electron Anomalies in the Old Quantum Theory. Hist Stud Phys Sci 15:81–106

Jordi Taltavull M (2018) The Uncertain Limits Between Classical and Quantum Physics: Optical Dispersion and Bohr's Atomic Model. Annalen der Physik 530.1800104. https://onlinelibrary. wiley.com/doi/full/10.1002/andp.201800104

---

[48] Sommerfeld to Nobel committee, 20 December 1917, NAS. Also in ASWB I. "Die Königl. Akademie der Wissenschaften zu Stockholm hat mir die Ehre erwiesen, mich zu einem Vorschlage über die Verteilung des physikalischen Nobelpreises für 1918 einzuladen. Indem ich dieser Einladung nachkomme, erlaube ich mir, als die zu krönende Entdeckung vorzuschlagen: die Quantentheorie von Max Planck [...]."

Kaiserfeld T (1993) When Theory Addresses Experiment: The Siegbahn-Sommerfeld Correspondence, 1917–1940. In: Lindqvist S (ed) Center on the Periphery. Historical Aspects of 20th-Century Swedish Physics. Science History Publications, Canton, MA, pp 306–324

Kragh H (1985) The Fine Structure of Hydrogen and the Gross Structure of the Physics Community, 1916–26. In: Historical Studies in the Physical Sciences 16, pp 67–125

Kragh H (2012) Niels Bohr and the Quantum Atom: The Bohr Model of Atomic Structure 1913–1925. Oxford University Press, Oxford

Leone M, Paoletti A, Robotti N (2004) A Simultaneous Discovery: The Case of Johannes Stark and Antonino Lo Surdo. Phys Perspect 6:271–294

Nielsen JR (ed) (1976) Niels Bohr: Collected Works. Vol. 3: The Correspondence Principle. (1918–1923). North Holland, Amsterdam

Nisio S (1973) The Formation of the Sommerfeld Quantum Theory of 1916. In: Japanese Studies in the History of Science 12, pp 39–78

Péres E (2009) Ehrenfest's Adiabatic Theory and the Old Quantum Theory, 1916–1918. Arch Hist Exact Sci 63:81–125

Robotti N (1986) The Hydrogen Spectroscopy and the Old Quantum Mechanics. Rivista di Storia della Scienza 3:45–102

Robotti N (1992) The Zeeman Effect in Hydrogen and the Old Quantum Theory. Rivista Internationale di Storia della Scienza 29:809–831

Rubinowicz A (1918) Bohrsche Frequenzbedingung und Erhaltung des Impulsmomentes. Physikalische Zeitschrift 19:441–445

Sommerfeld A (1913) Der Zeemaneffekt eines anisotrop gebundenen Elektrons und die Beobachtungen von Paschen-Back. Annalen der Physik 40:748–774

Sommerfeld A (1914) Zur Voigtschen Theorie des Zeeman-Effektes. In: Nachrichten von der Königlichen Gesellschaft der Wissenschaften zu Göttingen. Mathematischphysikalische Klasse, pp 207–229

Sommerfeld A (1915a) Zur Theorie der Balmerschen Serie. In: Sitzungs-berichte der mathematischphysikalischen Klasse der K. B. Akademie der Wissenschaften zu Müchen, pp 425–458

Sommerfeld A (1915b) Die Feinstruktur der Wasserstoff- und der Wasserstoff-ähnlichen Linien. In: Sitzungsberichte der mathematisch-physikalischen Klasse der K. B. Akademie der Wissenschaften zu München, pp 459–500

Sommerfeld A (1916a) Zur Quantentheorie der Spektrallinien. Annalen der Physik 51:1–94, 125–167

Sommerfeld A (1916b) Die Quantentheorie der Spektrallinien und die letzte Arbeit von Karl Schwarzschild. Die Umschau 20:941–946

Sommerfeld A (1916c) Zur Theorie des Zeemaneffektes der Wasserstofflinien, mit einem Anhang über den Starkeffekt. Physikalische Zeitschrift 17:491–507

Sommerfeld A (1916d) Zur Quantentheorie der Spektrallinien, Ergäzungen und Erweiterungen. In: Sitzungsberichte der mathematisch-physikalischen Klasse der K. B. Akademie der Wissenschaften zu München, pp 131–182

Sommerfeld A (1917) Die Drude'che Dispersionstheorie vom Standpunkte des Bohr'schen Modells und die Konstitution von H2, O2 und N2. Annalen der Physik 53:497–550

Stone AD (2005) Einstein's Unknown Insight and the Problem of Quantizing Chaos. Phys Today 1–7

Chapter 6
Synergy and Competition in the Quantum Network, 1919–1925

Abstract A few years prior to the advent of quantum mechanics in 1925 was a period of bustling activity in Munich, and in the new schools of the quantum network with Niels Bohr in Copenhagen and Max Born in Göttingen. The names of doctoral students to whom Sommerfeld entrusted one or the other research tasks on atomic spectra reads like a "who is who" of the quantum revolution: Wolfgang Pauli, Werner Heisenberg, Gregor Wentzel, to name only the more prominent among Arnold Sommerfeld's prodigies. Competition was a natural consequence of the growing quantum network. But there was also cooperation, e.g. when Sommerfeld spent half a year in 1922/23 as guest professor in the USA and Göttingen became a temporary home for some of his Munich students. Both Heisenberg and Pauli spent extended sojourns of their formative years in Munich, Göttingen, and Copenhagen. Sommerfeld himself entertained close relations to these and other centers, such as the spectroscopy group at the National Bureau of Standards.

Keywords Quantum network · Munich · Göttingen · Copenhagen · Max Born · Niels Bohr · Wolfgang Pauli · Werner Heisenberg · Gregor Wentzel · Otto Laporte · William Frederick Meggers · International Education Board · Rockefeller Foundation

With the end of the war and the ensuing revolutionary upheaval, most German physicists went through a period of uncertainty, hope, and depression. Sommerfeld found "everything unspeakable, miserable, and stupid," as he wrote to Einstein. "Our enemies are the biggest liars and scoundrels, we the biggest idiots. Not God but money rules the world."[1] However, anger about politics did not prevent him from pursuing physics. Over the course of 1918, X-ray spectroscopy became an issue of particular interest for Sommerfeld (Heilbron 1967). He turned to Siegbahn first, whenever he

---

[1] Sommerfeld to Einstein, 3 December 1918. AEA, Einstein. Also in ASWB I, doc. 295. "Ich höre von Kossel, dass Sie an die neue Zeit glauben und an ihr mitarbeiten wollen—Gott erhalte Ihnen Ihren Glauben! Ich finde alles unsagbar elend und blödsinnig. Unsere Feinde sind die grössten Lügner und Halunken, wir die grössten Schwachköpfe. Nicht Gott sondern das Geld regirt die Welt."

needed corroboration or refutation of one or another version of his model (Kaiserfeld 1993). Half a year after the war had ended, Siegbahn invited Sommerfeld to Lund. Sommerfeld perceived the invitation as a "dove of peace" and accepted it with great pleasure.[2]

Bohr, too, was eager to restore personal contacts with his foreign colleagues. When he sent Sommerfeld the second part of his treatise "On the Quantum Theory of Line-Spectra" in December 1918, he added: "Hopefully, scientific friends from different countries will be able to meet again in the coming year."[3] Sommerfeld appreciated "the extraordinarily liberal and faithful manner with which you acknowledge my own results and those of my students in your papers. Thereby colleagues in hostile countries, who otherwise tend to deny German accomplishments, will be forced to realize that, even during the war, German science could not be suppressed."[4] The latter remark reveals Sommerfeld's resentment toward the Entente, which had boycotted the work of German scientists (Schroeder-Gudehus 1966). When Bohr learned of Sommerfeld's sojourn in Lund, he extended an invitation to the German scientist to visit him in Copenhagen, which Sommerfeld was thrilled to accept.[5]

Both the German invitee and his Scandinavian hosts perceived these invitations as an extraordinary event. Sommerfeld spent four weeks in Scandinavia, from September 3 to October 1, 1919. Sommerfeld spent two weeks in Lund where his visit coincided with a conference of approximately 40 other physicists. Additional stops on his voyage included Copenhagen, Stockholm, and Uppsala; he was invited to give speeches and spent three or four days in each city. He was overwhelmed by the hospitality of his hosts. In his private letters to his wife and his daughter, he marveled at the welcoming spirit with which he was invited to participate in the family life of his hosts. "Bohr's young wife is very charming," he wrote to his nineteen-year-old daughter and continued:

> I also got to know Bohr's mother, a dear old lady. I could not abstain from telling the young Mrs. Bohr how glad I was to see Bohr in such good female hands, wife and mother. Both are concerned that he overworks himself and asked me to talk with his colleague Knudsen

---

[2]Sommerfeld to Siegbahn, 5 June 1919, ARSA, Siegbahn Papers. "[…] die mir Ihre freundliche Einladung als erste wirkliche Friedens-Taube erscheinen lassen. Ich werde sehr gern zu Ihnen kommen […]."

[3]Bohr to Sommerfeld, 26 December 1918, NBA, Bohr Papers. Also in ASWB I. "Hoffentlich werden wissenschaftliche Freunde aus den verschiedenen Ländern im kommenden Jahr einander wieder treffen können."

[4]Sommerfeld to Bohr, 5 February 1919. NBA, Bohr Papers. Also in ASWB II. "Ich danke Ihnen auch herzlich für die ausserordentlich liberale und gewissenhafte Art, mit der Sie meine und meiner Schüler Resultate in Ihren Arbeiten anerkennen. Dadurch werden wohl auch die Fachgenossen in den feindlichen Ländern, die sonst gern alle deutschen Leistungen unterschlagen möchten, gezwungen sein, einzusehen, dass sich die deutsche Wissenschaft selbst im Kriege nicht unterdrücken lässt."

[5]Bohr to Sommerfeld, 30 August 1919. NBA, Bohr Papers. Also in ASWB II. "[…] wir freuen uns Ihnen mitteilen zu können, dass Sie in einigen Tagen von 'Danmarks Naturvidenskabelige Samfund' […] eine officielle Einladung zu einem Besuche in Kopenhagen bekommen werden."

that he be less burdened. I did this, of course. Bohr is just like Einstein, only well groomed and more elegant […] I really became friends with him.[6]

He wrote Hilbert, "[My] reception everywhere has been very cordial, especially at Siegbahn's and Bohr's whom I now count as not only scientific but also personal friends. Bohr is a wonderful man."[7]

For Siegbahn and Bohr, eighteen and seventeen years younger than Sommerfeld respectively, the Munich theorist's visit took place at a crucial point in their careers. In December 1919, Johannes Rydberg, the head of the physics department at Lund University, died; in January 1920, Siegbahn was officially appointed as the new director, a position which he had unofficially held since Rydberg suffered a stroke in 1912 (Kaiserfeld 1993). Bohr was planning an "Institute for Work on Atomic Problems" around the same time, as Bohr's brother detailed in a letter to Sommerfeld in October 1919, requesting Sommerfeld's support for these plans.[8] So shortly after the war, there was more at stake than merely personal ambition and career. Bohr's new institute sought to establish Denmark as a scientific nation. As a neutral country, it hoped the stalement between the Entente and defunct German and Austro-Hungarian empires would be an opportunity for Denmark to compete scientifically on an international level.[9] Sommerfeld, for his part, regarded Bohr's plans as an instrument to counter the Entente's anti-German attitude. In his letter of support to the Carlsberg Funds, he emphasized the expected role of Bohr's institute as "an international workplace also for talented people from abroad, whose own countries can no longer offer the golden freedom of scientific work."[10] In the years to come, Sommerfeld's new friendships would prove beneficial to his own professional goals. With Siegbahn and Bohr as directors of their own institutes, Lund and Copenhagen emerged as important nodes in the expanding quantum network.

Meanwhile, the postwar situation in Germany precipitated a reassessment of theoretical physics. The universities reacted to the return of students from the trenches into the lecture halls by establishing crash lecture courses and new departments. At

---

[6]Sommerfeld to Margarethe Sommerfeld, 24 September 1919. Private Papers. Also in ASWB II, pp. 16–17. "Ganz entzückend ist die junge Frau Bohr […] Auch die Mutter von Bohr habe ich kennen gelernt, eine liebe alte Dame. Ich konnte mich nicht enthalten, der jungen Frau Bohr zu sagen, dass ich mich freute, Bohr in so guten weiblichen Händen, Frau u. Mutter, zu sehn. Beide sorgen sich, dass er sich überarbeitet, u. beide baten mich, bei seinem Collegen Knudsen dahin zu wirken, dass er entlastet würde. Das habe ich natürlich getan. Bohr ist ganz wie Einstein, nur viel besser gewaschen und viel feiner. […] Die Aufnahme in Kop. war wirklich herzlich, nicht nur von Seiten Bohrs, mit dem ich wirklich befreundet geworden bin […]."

[7]Sommerfeld to Hilbert, 25 September 1919, SUB, Cod. Ms. Hilbert 379. "Die Aufnahme war hier überall sehr herzlich, zumal bei Siegbahn und Bohr, mit denen ich nicht nur wissenschaftlich sondern auch persönlich befreundet bin. Bohr ist ein wundervoller Mensch."

[8]Bohr to Sommerfeld, 14 October 1919, NBA, Bohr Papers. Also in ASWB II. "[…] ein Institut für Arbeit in Atomfragen […]."

[9]See (Kojevnikov forthcoming 2020).

[10]Sommerfeld to Carlsberg Fund, 25 October 1919. NBA, Bohr Papers. Also in ASWB II. "[…] es sollte eine internationale Arbeitsstätte auch für Talente des Auslandes werden, denen die eigene Heimat nicht mehr die goldene Freiheit der wissenschaftlichen Arbeit gewähren kann."

universities where theoretical physics had not yet attained a firm standing, professor extraordinarius positions were promoted to chairs of theoretical physics (Jungnickel and McCormmach 1986). Lenz and Ewald, for example, who had returned to Munich in the winter of 1918–1919 to assist Sommerfeld as Privatdozenten during this extraordinary/crisis half year due to the war (*Kriegsnothalbjahr*), quickly received offers for new chairs in Hamburg and Stuttgart.[11] The situation in Munich also changed in other ways. In 1920, Wien succeeded Röntgen as director of the institute for experimental physics. With Röntgen in his seventies and frequently ill, the institute had been largely defunct during and after the war. "Wien has revived Röntgen's Sleeping Beauty institute," a relieved Sommerfeld expressed in February 1922 in a letter to Epstein. Epstein had accepted a professorship of theoretical physics at the California Institute of Technology in Pasadena where he established the first American node of the quantum network in 1921.[12] Sommerfeld was eager to resume collaboration with neighboring experimentalists, which had been so productive before the war. He also established close contacts with physical chemists who became natural allies in the quest for the atomic structure. "Presumably Herzfeld intends to habilitate here," Sommerfeld remarked in another letter to Epstein in October 1919.[13] Herzfeld was working as a "voluntary assistant" in the Vienna Radium Institute for his habilitation at the time. "In December 1919, the formalities were nearly completed," Herzfeld recalled in an unpublished autobiography, "when a letter arrived from Professor Sommerfeld in Munich, offering me a position as Privatdozent in theoretical physics and physical chemistry and a research assistantship with Professor Fajans."[14]

When Ewald and Lenz vacated their positions, Sommerfeld employed Adolph Kratzer and Gregor Wentzel as his new assistants. Kratzer had returned in October 1918 to Munich as a disabled veteran and assisted Sommerfeld throughout the early postwar period, although he was still a student. Sommerfeld particularly acknowledged Kratzer's help with *Atombau und Spektrallinien*, which appeared in September 1919. In February 1920, he completed a doctoral thesis on band spectra, which was supervised by both Sommerfeld and Lenz.[15] Wentzel had studied in Freiburg and

[11] Until 30 September 1920, Lenz was one of Sommerfeld's two assistants. He was awarded with the title and rank of professor extraordinarius on 11 November 1920, but shortly thereafter was offered the same title at the University of Rostock. In 1921, he was offered a new chair of theoretical physics in Hamburg (see Freddy Litten: Wilhelm Lenz—Kurzbiographie, http://litten.de/fulltext/lenz.htm, accessed 17 April 2020, available in German). Ewald succeeded Erwin Schrödinger as professor extraordinarius of theoretical physics at the Technische Hochschule Stuttgart on April 1, 1921. When he declined an offer from Münster in 1922 he was promoted to professor ordinarius (see https://www.archiv.ub.uni-stuttgart.de/veroeffentlichungen/Ewald_Paul-Peter.pdf, accessed 17 April 2020).

[12] Sommerfeld to Epstein, 12 February 1922, CA, Epstein Papers, 8.3. "Wien hat das Röntgensche Dornröschen-Institut zu neuem Leben erweckt."

[13] Sommerfeld to Epstein, 26 October 1919, CA, Epstein Papers, 8.3. "Voraussichtlich wird sich Herzfeld hier habilitieren."

[14] Autobiography and Family History, 1971. Typescript, CUA, Herzfeld Papers, box 2.

[15] Sommerfeld and Lenz, report to the faculty, 19 February 1920, UAM, OC I 46p. "Herr Kratzer (Kriegsinvalide, Kehlkopfschuss) ist seit Oktober 1918 mein Mitarbeiter und hat mir bei der Abfassung meines Buches über Atombau etc. wesentliche Dienste geleistet. Die vorliegende Arbeit

Greifswald before coming to Munich in 1920 and fell under Sommerfeld's spell. He, too, chose a topic from atomic theory (X-ray spectra) as a theme for his doctoral thesis.[16] "I again have excellent people," Sommerfeld rejoiced. "Wentzel has dissected the systematics of X-ray spectra in all their details [...] Kratzer is a master of band spectra." With Kossel from the Technical University, the reinstated institute of experimental physics, and an eager group of physical chemists around, Sommerfeld was satisfied with his local network. "The phenomenon, however, is Heisenberg, a third semester [student]," he marveled in a letter to Epstein of his recent prodigy student.[17] Besides Heisenberg, there was another student who displayed an "amazing precocity," as Sommerfeld described Pauli, who had completed his dissertation in the summer of 1921.[18] Pauli had come to Munich only two years before. "A first semester!" Sommerfeld was struck by Pauli's performance in January 1919. "His talent is many times better than even Debye's!!"[19]

With such unusual students in his lectures and seminars, Sommerfeld's tendency to involve them in his own research became a conspicuous trait of his nursery. Both Pauli and Heisenberg were confronted with the most advanced quantum problems long before they finished their studies with doctoral theses. Pauli, for example, corresponded with Landé "at the instigation" of Sommerfeld in December 1919—as a student in his third semester—on a recent publication of Landé's on the quantization of ring atoms, presenting him with a more elegant derivation.[20] Heisenberg, a year and a half younger than Pauli, also began participating in Sommerfeld's research at this stage of his studies. In March 1922, Sommerfeld announced a paper to Bohr on the anomalous Zeeman effect, which he had co-authored with Heisenberg. "You will

schliesst teils an ein Seminar vom Sommersemester 1919, teils an eine Untersuchung über Bandenspektren von Dr. Lenz an. Letzterer berichtet in dem folgenden Referat über Einzelheiten der Kratzerschen Arbeit [...]."

[16] Sommerfeld, report to the faculty, 21 June 1921, UAM, OC–I–47p. "Herr G. Wentzel hat durch seine Arbeit die Systematik der Röntgenspektren ein gutes Stück vorwärts gebracht. [...] Die vielen bis vor kurzem noch rätselhaften Linien der L-Serie sind jetzt vollständig in bestimmte Anfangs- und Endniveaus eingeordnet, mit solcher Sicherheit, dass Herr W. das Ergebnis der parallel laufenden Messungen von Coster vielfach vorhersagen konnte [...]."

[17] Sommerfeld to Epstein, 12 February 1922, CA, Epstein Papers, 8.3. "In der Münchener Physik herrscht reges Leben. [...] Wentzel hat die Systematik der Röntgenstr. bis in alle Einzelheiten geklärt [...] Kratzer beherrscht die Banden. Das Phänomen aber ist Heisenberg, ein 3tes Semester, der die Modelltheorie der Zeemaneffekte und der multiplen Terme erschlagen hat [...]."

[18] Sommerfeld, report to the faculty, 21 July 1921, UAM, OC I 47p. "Der hochbegabte junge Verfasser dieser Arbeit ist von erstaunlicher Frühreife. Als er nach München kam, war er schon im vollen Besitz der mathematischen und mathematisch-physikalischen Methoden. Er brachte eine fertige Arbeit zur allgemeinen Relativitätstheorie mit, die sofort Einsteins Aufmerksamkeit und Bewunderung erregte. Eine mir sehr wertvolle Leistung war sodann die Abfassung eines grossen Artikels über Relativitätstheorie für die mathematische Enzyklopädie; auch bei meinem Buch über "Atombau [...]" hat er mir erfolgreich assistiert."

[19] Sommerfeld to Geitler, 14 January 1919. Private Papers. Also in ASWB II. "Ein erstes Semester! Seine Begabung geht selbst über die von Debye um ein Vielfaches hinaus!".

[20] Pauli to Landé, 18 December 1919. In WPWB I. "Auf Veranlassung von Herrn Geheimrat Sommerfeld habe ich Ihre Arbeit über die Berücksichtigung der Wechselwirkung von Kreisbahnen bei Ihrer Quantelung mit Hilfe der Adiabatenhypothese genauer studiert [...]."

see in the paper that we interpret the term multiplicity as a result of a magnetic inter-action," he contrasted Heisenberg's effort from previous interpretations including Bohr's own explanation.

> Heisenberg is a student in his third semester and enormously gifted. I could not tame his eagerness for publication any longer and regard his results as so important that I agreed publication was necessary even if the method of derivation is probably not yet definitive.[21]

Heisenberg's interpretation became known as the "rump model" and sparked debates about the anomalous Zeeman effect (Cassidy 1979; Seth 2008).

Sommerfeld involved not only his exceptional students in his quantum research. One of his older students, Josef Krönert, had studied with him before the First World War. Throughout the war, Krönert expressed an unabated interest in physics, and Sommerfeld sent him recent research papers to his various sites of deployment. In August 1916, for example, when he was close to the so-called witches' cauldron of Verdun, Krönert thanked Sommerfeld for sending him Ewald's and Landé's arti-cles.[22] As soon as he returned to Munich after the war, Sommerfeld asked him to scrutinize the available data concerning the anomalous Zeeman effect. "The work of Mr. Krönert contributes appreciably to my own studies on spectra and the Zeeman effect in various regards," Sommerfeld commented on this work, which he presented to the faculty as Krönert's doctoral thesis, although its results were "for the time the-oretically incomprehensible, like all the other processes in the anomalous Zeeman effect." But he appreciated that "the work is done with devotion and reflects a good intuition in this momentarily still rather dark specialty."[23] Another student whose study was interrupted by the war was Erwin Fues. When he returned to the univer-sity after the war, Sommerfeld assigned him a spectroscopic theme as subject for a doctoral thesis. In 1919, Sommerfeld and Kossel had postulated a "spectroscopic displacement law" that stated there was a similarity in the spark spectrum of an atom with the arc spectrum of the atom in the preceding row of the periodic table. In other words, the spectrum of a singly ionized atom with element number Z was similar to that of the neutral atom with Z-1. Fues had the task of checking this conjecture.[24]

---

[21]Sommerfeld to Bohr, 25 March 1922, NBA, Bohr Papers. Also in ASWB II. "Sie werden daraus sehen, dass wir den Ursprung der Termmultiplicitäten in's Magnetische legen […] Heisenberg ist Student im 3. Semester und ungeheuer begabt. Ich konnte seinen Publikationseifer nicht länger zügeln und finde seine Resultate so wichtig, dass ich ihrer Veröffentlichung zustimmte, trotzdem die Form der Ableitung noch nicht die definitive sein dürfte."

[22]Krönert to Sommerfeld, 6 July 1916, DM, NL 89, 059.

[23]Sommerfeld, report to the faculty, 20 February 1920, UAM, OC I 46 p. "Die Arbeit des Herrn Krönert bedeutet nach verschiedenen Richtungen einen schätzenswerten Beitrag zu meinen eigenen Studien über Spektren und Zeemaneffekt […] Natürlich ist sie theoretisch zunächst unverständlich, ebenso wie die ganzen sonstigen Vorgänge im anomalen Zeemaneffekt. […] Die Arbeit ist mit Hingabe gemacht und zeigt ein gutes Orientierungsvermögen in dem zur Zeit noch recht dunklem Gebiet."

[24]Sommerfeld, report to the faculty, 18 December 1919, UAM, OC I 46p. "Herr Fues hatte die Aufgabe, einen Satz ("spektroskopischer Verschiebungssatz") näher zu prüfen, den Kossel und ich kürzlich aufgestellt haben. Nach diesem Satz besteht ein enger Zusammenhang zwischen dem Funkenspektrum eines Elements und dem Bogenspektrum des im periodischen System vorherge-

By the early 1920s, Sommerfeld's institute had become a place where talented physics students competed to display their skills. The first opportunity to attract the professor's attention came with Sommerfeld's lectures. "The most important part of the lecture was the exercises connected with the lectures," Heisenberg recalled. "Of course, the professor tried to get some impression of how good the student was in solving these problems." The next stage was the seminar. In Heisenberg's case, this was the first step; when he approached Sommerfeld with the request to attend the seminar immediately, he recalled that Sommerfeld was very friendly and said,

> "All right, you have an interest in mathematics; it may be that you know something; it may be that you know nothing; we will see. All right, you come to the seminar, and we will see what you can do." So I came to Sommerfeld, and already four weeks after I attended his seminar he gave me a problem, which was, of course, very nice.[25]

In general, Sommerfeld used the seminar as a sounding board for discussions on contemporary research. "There, of course, Sommerfeld tried to get information about the most recent developments himself. That he did by turning to a student and saying, 'Now here you have the paper of Mr. Kramers. You give a talk at the seminar next week and explain to us what Kramers actually means by his paper and what you think about it.'"[26] Other Sommerfeld disciples from this period recalled similar experiences.[27] From 1919 to 1922, at least ten doctoral students graduated from Sommerfeld's nursery, five of them with quantum themes (Fues, Krönert, Kratzer, Pauli, Wentzel; see Table A.2). Two of them, Kratzer and Wentzel, proceeded with a habilitation thesis and advanced to Privatdozenten (Table A.3). Fues did the same as Ewald's assistant in Stuttgart and Pauli as Lenz's assistant in Hamburg (after sojourns with Born in Göttingen and Bohr in Copenhagen, respectively). By comparison, only three doctoral theses and one habilitation were accomplished in the same period at Planck's institute in Berlin (Beck 2009, 88–89).

The thriving quantum environment of Munich can also be seen in lecture topics of Sommerfeld and his Privatdozent lectures (Table A.4): in the summer semester of 1919, Lenz lectured on *Quantentheorie* (quantum theory); during the winter semesters of 1919–20 and 1920–21, Sommerfeld dedicated popular lectures to *Atombau und Spektrallinien* (atomic structure and spectral lines) and *Theorie der Spektrallinien auf Grund des Bohr'schen Atommodelles* (theory of spectral lines based on Bohr's atomic model); in the winter semester of 1921–22, Herzfeld lectured on *Quantenmechanik der Atommodelle* (quantum mechanics of atomic models). The

---

henden Elements [...] Die Arbeit ist sehr sorgfältig ausgeführt [...] Sie hat das gestellte aktuelle Problem sachgemäss gelöst und wird auch für die allgemeine Methodik der Serienberechnung wertvoll sein."

[25] Interview with Heisenberg by Kuhn and Heilbron, 30 November 1962, https://www.aip.org/history-programs/niels-bohr-library/oral-histories/4661-1, accessed 17 April 2020.

[26] Interview with Heisenberg by Kuhn and Heilbron, 30 November 1962, https://www.aip.org/history-programs/niels-bohr-library/oral-histories/4661-1, accessed 17 April 2020.

[27] Interviews with Wentzel, 3–5 February 1964, and Otto Laporte, 29 January 1964, by Kuhn, https://www.aip.org/history-programs/niels-bohr-library/oral-histories/4953; https://www.aip.org/history-programs/niels-bohr-library/oral-histories/4731-1, accessed 17 April 2020.

Munich Wednesday Colloquium (which was now held mostly on Fridays) was the stage for presenting doctoral and habilitation theses in Sommerfeld's institute and for current atomic and quantum research from elsewhere, such as Walter Gerlach's presentation on May 26, 1922 about recent experiments on *Atomstrahlen im Magnetfeld* (atomic beams in a magnetic field), better known as the Stern-Gerlach experiment.[28] In the same period from 1919 to 1922, Sommerfeld published three editions of *Atombau und Spektrallinien*, an effort that required the help of his advanced students, whom Sommerfeld acknowledged.[29] "I have slogged away, particularly at the new edition of my book and am now ripe for a holiday," Sommerfeld wrote to Einstein after the summer semester in 1921. "I have made four doctors (among them Pauli) and one Privatdozent (Kratzer). All this takes a lot of effort."[30]

Thus quantum theory rapidly developed into a desirable field of study with many opportunities for theoretical physicists beginning their careers. The rise of quantum activities across the country, provoked the emergence of rivalries between theorists in Munich as theorists elsewhere. When Born succeeded Debye, who had accepted the directorship of the physics institute at the Swiss Federal Institute of Technology Zurich (Eidgenössische Technische Hochschule Zürich, ETH), Göttingen turned into a rival node of the quantum network. "Our perturbation method (basically the same as that of Bohr and Kramers) seems to lead further than Kratzer's," Born wrote to Sommerfeld in May 1922. "In other regards, too, I let my people quantize [*quanteln*] in order to compete with you a little."[31] This friendly competition, however, could turn into fierce conflicts over priorities in one quantum field or another. Born, for example, considered the perturbation theory as his particular specialty, thereby competing with Epstein, who had earlier started to employ this field as a resource for quantum theory. Born wrote to Sommerfeld on one occasion:

> If you talk to Epstein in Pasadena and he is angry with me, then ask him to show you the very unfriendly letter that he wrote to me because he felt that Pauli's and my own work on perturbation theory deprived him of his supposed birthright to the field. Tell him also that I do not answer such letters, but that I am not resentful about his discourtesy (mildly put) and would like to be friends with him, if he so desires, in an appropriate form. By the way, we are actually further advanced in the questions of perturbation quantization than he.[32]

---

[28] Physikalisches Mittwoch-Colloquium, DM, 1997–5115.

[29] See (Eckert 2013).

[30] Sommerfeld to Einstein, 10 August 1921, AEA, Einstein Papers. "Ich habe viel geschuftet, besonders an der neuen Aufl. meines Buches, u. bin jetzt ferienreif. Habe in diesem Semester 4 Doctoren (unter ihnen Pauli) und 1 Privatdocenten (Kratzer) gemacht. Das kostet alles Schweiß."

[31] Born to Sommerfeld, 13 May 1922, DM, HS 1977–28/A,34. Also in ASWB II. "Unsere Störungsmethode (übrigens wesentlich dieselbe, die Bohr und Kramers haben) scheint weiter zu führen als Kratzer. Auch sonst lasse ich jetzt meine Leute 'quanteln', um Ihnen ein wenig Konkurrenz zu machen."

[32] Born to Sommerfeld, 5 January 1923, DM, HS 1977–28/A,34. Also in ASWB II. "Wenn Sie Epstein in Pasadena sprechen und er etwa auf mich schimpft, so sagen Sie ihm, er solle Ihnen den recht unfreundlichen Brief zeigen, den er mir geschrieben hat, weil er sich durch Paulis und meine Störungsarbeit in seinem Erstgeburtsrecht benachteiligt glaubte. Sagen Sie ihm ferner, daß ich solche Briefe nicht beantworte, ihm aber seine Unhöflichkeit (eine milde Bezeichnung) nicht

Other rivalries arose between experimentalists and theorists: Paschen's disciple Back, for example, felt disenfranchised by Landé, who benefited from his experiments on the anomalous Zeeman effect and rushed to publication without regard to Back's slower pace of presenting his results in a coherent manner.[33] In Sommerfeld's institute, Wentzel competed with one of Siegbahn's students (Dirk Coster) who approached the interpretation of X-ray data from an experimental point of view.[34]

Competition was only one consequence of the growing quantum network. It also witnessed an increase in collaborative spirit due to a shared devotion to a common issue. Sommerfeld benefited from this synergy when Born agreed to supervise four advanced students on his behalf during the winter semester of 1922–23, which Sommerfeld spent in the United States as guest professor at the University of Wisconsin in Madison. The rise of quantum topics as themes of doctoral thesis shows up in a letter in which Born told Sommerfeld how disadvantaged some of the impacted students felt in this situation:

Starting with Mr. Fisher, he is not to blame for his bad luck, because a Dutch man, Niessen, has just published an article in *Physica* which makes it clear that he (apparently independently from Pauli) has calculated the H_2^+, including the band spectrum. This renders Fisher's work obsolete. Now he has asked me for a new theme. I told him that I cannot suggest one for the time being; he should participate in the seminar, perhaps he will find an appropriate task for himself.[35]

Born felt exhausted. Apart from Sommerfeld's students, he had his own group of advanced students to advise. But of the Munich students, Born was particularly impressed by Heisenberg. "I have become fond of Heisenberg," he confided to Sommerfeld. He proposed that Heisenberg should begin his career as his assistant after finishing the doctoral degree in Munich.

When I asked him about his plans he responded: 'This I do not get to decide! This is determined by Sommerfeld!' You are therefore his self-selected guardian and I have to stick with you if I want to lure him away to Göttingen.[36]

Competition and synergy were also characteristic features of Sommerfeld's relationship with Bohr. By 1922, three of Sommerfeld's disciples chose Bohr's institute

nachtrage und gern gut Freund mit ihm sein will, wenn er es in angemessener Form wünscht. Übrigens sind wir in den Fragen der Störungsquantelung doch weiter als er."

[33] Back to Sommerfeld, 7 June 1921, DM, HS 1977–28/A,8. See also (Forman 1970).

[34] Coster to Sommerfeld, 23 May 1921, DM, NL 89, 006.

[35] Born to Sommerfeld, 5 January 1923, DM, HS 1977–28/A,34. Also in ASWB II. "Da ist erst Herr Fisher; dieser kann nichts für sein Pech, denn ein Holländer Niessen hat in der Physika einen Aufsatz publiziert, woraus hervorgeht, daß er (anscheinend unabhängig von Pauli) das H_2^+ berechnet hat, und zwar einschließlich des Bandenspektrums. Damit ist wohl Fishers Arbeit erledigt. Nun will er ein Thema von mir. Ich habe ihm gesagt, daß ich ihm vorläufig keins geben kann; er soll im Seminar mitarbeiten, vielleicht stößt er dabei von selbst auf eine brauchbare Aufgabe."

[36] Born to Sommerfeld, 5 January 1923, DM, HS 1977–28/A,34. Also in ASWB II. "Heisenberg habe ich sehr lieb gewonnen [...] Heisenberg will im Sommer bei Ihnen in München promovieren. Als ich ihn fragte, was er nachher vorhabe, antwortete er: 'Das habe ich doch nicht zu entscheiden! Das bestimmt Sommerfeld!' Sie sind also sein selbsterkorener Vormund, und ich muss mich an Sie halten, wenn ich Heisenberg nach Göttingen ziehen will."

for extended research sojourns (Rubinowicz, Landé, and Pauli). Events like the "Bohr Festival," a series of lectures presented by Bohr at Göttingen in June 1922, further strengthened the quantum network.[37] Sommerfeld used Bohr's Göttingen sojourn as an opportunity to introduce Heisenberg to the assembled elite of quantum theorists.[38] If Bohr felt "scientifically very lonely," as he confessed to Sommerfeld in April 1922,[39] the attention he experienced during his Göttingen lectures must have changed this attitude. At the same time, the Göttingen event—and its label as a "festival" to which Munich, Copenhagen and Göttingen payed due tribute—marked the recognition of Born's institute as the third major node of the quantum network.

Sommerfeld's invitation from the University of Wisconsin in Madison further extended the network.[40] Sommerfeld's appointment as the Karl Schurz Professor—a chair that was founded in 1910 to foster German-American relations—was noted as a remarkable event so shortly after the end of the First World War. "German Scientist Coming," the *New York Times* reported on August 6, 1922. But it was not only the political context that sparked curiosity. "Professor Sommerfeld is expected to give a course on atomic structure," the newspaper further reported.[41] Advances in physics had already resulted in invitations to overseas professors from several American universities, such as Einstein, Curie, and Lorentz (Sopka 1980, Appendix II). With Sommerfeld, the invitation revealed a particular interest in the recent European development of quantum physics. At the time of Sommerfeld's arrival in September 1922, the English translation of *Atombau und Spektrallinien* was still in the making, but Sommerfeld's book's fame preceded the English translation. The third edition had just appeared, and news of Sommerfeld's arrival resulted in a flood of invitations to lecture on the subject of his book throughout the United States. While he was based in Madison from September 1922 to January 1923, he visited Evanston, Milwaukee, Minneapolis, Ann Arbor, Urbana, and St. Louis. Subsequently, he traveled to the California Institute of Technology in Pasadena and the University of California, Berkeley, where he spent two weeks at each. On his way to the West Coast, Sommerfeld stopped in Kansas, and on his return he visited Denver and Ames for another couple of lectures. In March 1923, Sommerfeld concluded his American journey with a round trip through Eastern states, lecturing in Washington D.C., Schenectady, Cambridge (Massachusetts), Ithaca, and New York City. Altogether, Sommerfeld lectured at seventeen locations during his six-month sojourn in the United States.[42] Shortly after Sommerfeld returned, the American affiliations with the European quan-

---

[37] See (Schirrmacher 2019).

[38] Sommerfeld to Kramers, 1 June 1922, NBA, Bohr Papers.

[39] Bohr to Sommerfeld, 30 April 1922, DM, HS 1977–28/A,28. Also in ASWB II. "In die letzteren Jahren habe ich mich oft wissenschaftlich sehr einsam gefühlt unter dem Eindruck dass meine Bestrebungen, nach besten Vermögen die Principien der Quantentheorie systematisch zu entwickelen, mit sehr wenig Verständniss aufgenommen worden ist."

[40] Raymond Thayer Birge to Sommerfeld, 5 July 1922, DM, NL 89, 019, Mappe 4,1.

[41] German Scientist Coming. Professor A. Sommerfeld to Give Courses in Wisconsin University. *The New York Times*, 6 August 1922.

[42] This survey is based primarily on letters of invitation, preserved in DM, NL 89, and Sommerfeld's correspondence with his wife.

tum network were further strengthened by other invitations. Bohr visited the United States in the fall of 1923, where he presented lectures and participated in colloquia at Amherst College, Yale University, and Harvard University and, as well as at a meeting of the American Physical Society in Chicago. Within the subsequent two years, Oskar Klein, Siegbahn, and Born accepted similar invitations from American universities (Sopka 1980, Appendix II).

For ten days during his American residency, Sommerfeld collaborated with the spectroscopy department of the National Bureau of Standards in Washington, D.C., headed by William Frederick Meggers, on the interpretation of complex spectra. During a previous visit to Madrid, Sommerfeld had learned about Miguel Catalan's discovery of multiplets in the spectrum of manganese (Sanchez-Ron 1983). He realized that he could order these complex spectral patterns by means of the "inner quantum number," a concept he had introduced in 1920. This extension exposed a new potential of quantum theory as a tool for practical spectroscopy. The complex spectra of atoms with a myriad of lines that seemed inaccessible to further analysis, like that of iron, now became amenable to quantum theory. "Practical spectrographic analysis," as Meggers and his collaborators from the Bureau of Standards titled the article in 1922, thus was transformed from an empirical art into a science. Sommerfeld's ten-day visit at the bureau in March 1923 was the prelude to further collaborations between the Munich theorists and American practical spectroscopists and astrophysicists. Shortly after this visit, Sommerfeld informed Meggers that "we have discovered some multiplets in the spectrum of titanium and vanadium (from which we received the Zeeman effects from Pasadena). Therefore, the comparison with Fe would be very interesting."[43] By "we" Sommerfeld meant himself and his advanced student Otto Laporte, and by "Pasadena" he referred to the astrophysicists from the Mount Wilson observatory. A year later, Laporte wrote his dissertation on the spectrum of iron, a task which "was still considered hopelessly complicated a few years ago," as Sommerfeld argued in his report to the faculty.[44]

This German-American exchange happened at a time when Germany was still officially excluded from international scientific collaborations. In 1924, the German Foreign Office supported Sommerfeld's proposal of an exchange program with the Massachusetts Institute of Technology "as a low-key measure to bring young scholars of both countries together in order to foster the convergence of German and American institutes."[45] In the course of this program, the first American research students

---

[43]Sommerfeld to Meggers, 30 June 1923, AIP, Meggers Papers. "Ich möchte Sie nun doch bitten, mir das Fe-Spektrum des Mt. Wilson (zusammen mit dem Brief von Babcock) zuzuschicken. Nachdem Landé (Zeitschrift für Physik, Bd. 15, S. 189) die Zeemaneffekte für beliebige Multipletts theoretisch-empirisch erklärt hat, haben wir im Titan- und Vanadium-Spektrum (von denen wir die Zeeman-Effekte von Pasadena erhalten haben) einige Multipletts herausgefunden. Der Vergleich mit Fe wäre daher sehr interessant."

[44]Sommerfeld's report to the faculty, 26 July 1924, UAM, OC–I–50p. "Das Eisenspektrum galt noch bis vor wenigen Jahren als hoffnungslos kompliziert." In 1925, Laporte spent one year doing a fellowship as a theoretical analyst in Meggers's department at the National Bureau of Standards. On the iron spectrum, see (Grotrian 1924).

[45]Foreign Office to Sommerfeld, 5 April 1924, DM, NL 89, 030, "Die Ausführungen Euer Hochwohlgeboren im Schreiben vom 18. v. M. haben mich in der Tat überzeugt, daß der von

(Ernst and Victor Guillemin) went to Munich to study with Sommerfeld. At the same time, Catalan was granted a fellowship by the International Education Board of the Rockefeller Foundation to collaborate with Sommerfeld's student Karl Bechert on complex spectra. Bechert pursued this work with a doctoral dissertation on the spectrum of nickel. "The work of Mr. Bechert is in line with the investigations that are particularly cultivated in my institute, namely the disentanglement of complex spectra," Sommerfeld outlined in his report of Bechert's dissertation to the faculty. "After the iron spectrum had been studied by Laporte and that of cobalt by Catalan and Bechert, the nickel spectrum lent itself as a subject of particular interest."[46] With the analysis of these transition metals, Sommerfeld's theorists served not only quantum theory, they once again contributed to the field of practical spectroscopy. "I was pleased to learn what Catalan and Bechert are doing in your laboratory although the same will not please Walters," Meggers rejoiced of the combination of theory and practice in Sommerfeld's institute, although it gave rise to a rivalry with one of his own collaborators.[47]

In the course of these applications, Sommerfeld's quantum approach became more and more empirical. He became critical of some of his own earlier approaches, which were based on the construction of one or another model about the motion of electrons around atomic nuclei. With the effort to account for the spectral patterns displayed with the anomalous Zeeman effect, Sommerfeld felt reassured of his "whole-number" quantum approach. At the same time, he was critical about the underlying model assumptions. "I attach particular importance to the introduction of inner quantum numbers," he explained in 1922 in the preface to the third edition of *Atombau und Spektrallinien*. Regarding the anomalous Zeeman effect, he wrote, "doubts may arise only as far as the interpretation in terms of a model is concerned" (Sommerfeld 1922, preface).[48] With the extension of the theory to the complex spectra, the demise of model-based approaches even became praised as a virtue. Pauli admired Sommerfeld's approach in the fourth edition of *Atombau und Spektrallinien*:

Ihnen eingeleitete Assistentenaustausch mit dem Massachusetts Institute of Technology als unauffällige Art, jüngere Wissenschaftler beider Länder aneinander zu bringen und damit letzten Endes eine Annäherung deutscher und amerikanischer Institute zu fördern Unterstützung verdient."

[46]Sommerfeld, report to the faculty, 18 May 1925, UAM, OC–I–51p. "Die Arbeit des Herrn Bechert liegt in der Linie der in meinem Institut besonders gepflegten Untersuchungen über die Entwirrung komplizierter Spektren. Nachdem das Eisenspektrum von Laporte und das Cobaltspektrum von Catalan und Bechert studiert war, bot das Nickelspektrum ein besonderes Interesse."

[47]Meggers to Sommerfeld, 9 February 1925. AIP, Meggers Papers.

[48]"Besonderen Wert lege ich auf die Einführung der inneren Quantenzahlen (6. Kap., § 5) und auf die Systematik der anomalen Zeemaneffekte (6. Kap., § 7). Die hier herrschenden durchgreifenden Regelmäßigkeiten sind zunächst empirischer Natur; ihr ganzzahliger Charakter verlangt aber von Anfang an nach quantentheoretischer Einkleidung. Diese Einkleidung ist ebenso wie die Regelmäßigkeiten selbst völlig gesichert und eindeutig; sie hat sich schon jetzt für die praktische Spektroskopie vielfach als fruchtbar und anregend erwiesen. Zweifel können nur in bezug auf die modellmäßige Deutung entstehen."

When we speak in terms of models, we use a language that is not sufficiently adequate to
the simplicity and beauty of the quantum world. For this reason, I found it so nice that your
presentation of the complex structure is entirely free of all model prejudices.[49]

On the eve of quantum mechanics, therefore, the prevailing method in Sommerfeld's nursery seems to have changed from a model-based deductive approach to an
inductive method, searching for quantum regularities without "model prejudices."
However, Pauli's opinion should not lead us to assume a common trait in Sommerfeld's school. In his own research, Sommerfeld hardly followed an approach
that could be labelled as exclusively one or the other. In 1923, when he seemed to
have a preference for the inductive approach in his collaboration with the spectroscopists from the Mount Wilson observatory and the National Bureau of Standards,
he also published a theory about a model of the helium atom that belied the demise
of "model prejudices" by making assumptions (like coplanar electron motions in
opposite senses) that he had criticized shortly before in other problems (Sommerfeld 1923). When Sommerfeld reflected on the "Principles of Quantum Theory and
Bohr's Atomic Model" in 1924, he argued: "as a last resort every fundamental theory
in physics must proceed deductively," but that quantum theory was "not yet ripe"
for this approach. He characterized Bohr's procedure to arrive at quantum laws by
means of the correspondence principle as "inductive" and praised it as a successful
heuristic instrument (Sommerfeld 1924):

> The magic power of the correspondence principle has generally proven to be of value in
> selection rules for quantum numbers and in the series and band spectra. The principle has
> become the guide for all recent discoveries of Bohr and his disciples. Yet I still cannot regard
> it as satisfactory because of its mixture of quantum theoretical and classical points of view.
> I tend to consider the correspondence principle as a particularly valuable consequence of a
> future complete quantum theory, but not as its foundation.[50]

This passage accurately reflects Sommerfeld's attitude toward Bohr's correspondence principle. His opinion is often described as hostile, but should rather be
regarded as ambivalent; Sommerfeld himself considered the principle useful, but
not satisfactory. Whether Sommerfeld's dissatisfaction was based only on conceptual grounds is difficult to decide. It was probably also accompanied by a feeling of

---

[49]Pauli to Sommerfeld, 6 December 1924, DM, HS 1977–28/A,254. Also in ASWB II. "Man hat
jetzt stark den Eindruck bei allen Modellen, wir sprechen da eine Sprache, die der Einfachheit und
Schönheit der Quantenwelt nicht genügend adäquat ist. Deswegen fand ich es so schön, daß Ihre
Darstellung der Komplexstruktur von allen Modell-Vorurteilen ganz frei ist."

[50]"Jede grundlegende physikalische Theorie muß letztes Endes deduktiv vorgehen. […] Man kann
sagen, daß die Quantentheorie noch nicht reif ist für eine rein deduktive Darstellung […] Demge-
genüber sucht Bohr in seinem Korrespondenzprinzip die Quantentheorie eng an die klassische
Strahlungstheorie anzuschließen. Er geht möglichst induktiv und physikalisch vor, indem er schrit-
tweise jede Quantenzahl einer Bewegungsperiode zuordnet. Die Zauberkraft des Korresponden-
zprinzips hat sich allgemein bewährt, bei den Auswahlregeln der Quantenzahlen, in den Serien-
und Bandenspektren. Das Prinzip ist der Leitfaden geworden für alle neueren Entdeckungen Bohrs
und seiner Schüler. Trotzdem kann ich es. nicht als endgültig befriedigend ansehen, schon wegen
seiner Mischung quantentheoretischer und klassischer Gesichtspunkte. Ich möchte das Korrespondenzprinzip als eine besonders wichtige Folge der zukünftigen, vervollständigten Quantentheorie,
aber nicht als deren Grundlegung ansehen."

competition, as Sommerfeld felt that the centrality of his nursery was fading away and Bohr's institute—with his former prodigy students, Pauli and Heisenberg, as Copenhagen converts–was assuming the leading role.

However, with the advent of wave mechanics, there arose a new opportunity to attain centrality in the quantum network. The almost simultaneous birth of quantum mechanics in two varieties, wave and matrix mechanics, once more made synergy and competition paramount features between the nodes of the quantum network.

References

Beck LF (ed) (2009) Max Planck und die Max-Planck-Gesellschaft. Zum 150. Geburtstag am 23. April 2008 aus den Quellen zusammengestellt. Archiv zur Geschichte der Max-Planck-Gesellschaft, Berlin

Cassidy D (1979) Heisenberg's First Core Model of the Atom: The Formation of a Professional Style. Hist Stud Phys Sci 10:187–224

Eckert M (2013) Die Bohr-Sommerfeldsche Atomtheorie: Sommerfelds Erweiterung des Bohrschen Atommodells 1915/16. Klassische Texte der Wissenschaft. Springer, Berlin

Forman P (1970) Alfred Landé and the Anomalous Zeeman Effect, 1919–1921. HSPS 2:153–261

Grotrian W (1924) Die Entwirrung komplizierter Spektren, insbesondere des Eisenspektrums. Die Naturwissenschaften 12:945–955

Heilbron JL (1967) The Kossel-Sommerfeld Theory and the Ring Atom. Isis 58:450–485

Jungnickel C, McCormmach R (1986) Intellectual Mastery of Nature: Theoretical Physics from Ohm to Einstein, vol 2. The University of Chicago Press, Chicago

Kaiserfeld T (1993) When Theory Addresses Experiment: The Siegbahn-Sommerfeld Correspondence, 1917–1940. In: Lindqvist S (ed) Center on the Periphery. Historical Aspects of 20th-Century Swedish Physics. Science History Publications, Canton, MA, pp 306–324

Kojevnikov A (forthcoming 2020) KnabenPhysik, or the Birth of Quantum Mechanics from a Postdoctoral Perspective. Springer Briefs in History of Science and Technology. Springer Nature Switzerland AG, Cham

Sanchez-Ron JM (1983) Documentos para una historia de la fisica moderna en Espania: Arnold Sommerfeld, Miguel A. Catalan, Angel del Campo y Blas Cabrera. Llull 5:97–109

Schirrmacher A (2019) Establishing Quantum Physics in Göttingen: David Hilbert, Max Born, and Peter Debye in Context, 1900–1926. Springer Briefs in History of Science and Technology. Springer Nature Switzerland AG, Cham

Schroeder-Gudehus B (1966) Deutsche Wissenschaft und internationale Zusammenarbeit 1914–1928: Ein Beitrag zum Studium kultureller Beziehungen in politischen Krisenzeiten. Dumaret and Golay, Geneva

Seth S (2008) Crafting the Quantum: Arnold Sommerfeld and the Older Quantum Theory. Stud Hist Philos Sci 39:335–348

Sommerfeld A (1922) Atombau und Spektrallinien, 3rd edn. Vieweg, Braunschweig

Sommerfeld A (1923) The Model of the Neutral Helium Atom. J Opt Soc Am 7:509–515

Sommerfeld A (1924) Grundlagen der Quantentheorie und des Bohrschen Atommodelles. Die Naturwissenschaften 12:1047–1049

Sopka KR (1980) Quantum Physics in America. Arno Press, New York

Chapter 7
Wave Mechanics—A Pet Subject of the Sommerfeld School, 1926–1928

Abstract Arnold Sommerfeld gave only a lukewarm response to "matrix mechanics," as Werner Heisenberg's quantum achievement of 1925 was called; in contrast, his reaction to Erwin Schrödinger's "wave mechanics" was most enthusiastic. Schrödinger's equation proved to be a new tool for solving old problems. Sommerfeld declared this to be the pet research subject for his seminar and for some of his doctoral students, for example, Albrecht Unsöld and Hans Bethe. Sommerfeld also arranged for his students Walter Heitler and Fritz London to spend research sojourns as Rockefeller fellows with Schrödinger in Zurich. The quantum network was further expanded when Werner Heisenberg and Wolfgang Pauli became professors of theoretical physics at the University of Leipzig in 1927, and at the ETH Zurich in 1928, respectively. At the same time, Sommerfeld had revived the classical free-electron-gas model by replacing the Maxwell-Boltzmann statistics with the new Fermi-Dirac-statistics. This semi-classical electron theory of metals paved the way for the modern quantum theory of solids. For a number of years, Sommerfeld's advanced students and American research fellows began their careers with a research problem in this specialty. By the end of the 1920s, quantum mechanics had become a target of opportunity for a new generation of doctoral students who were starting their careers in theoretical physics.

Keywords Werner Schödinger · Wave mechanics · Albrecht Unsöld · Hans Bethe · Walter Heitler · Fritz London · Electron theory of metals · Linus Pauling · Leipzig · Zurich

Early publications on quantum mechanics by Heisenberg, Born, Jordan, and Dirac had no immediate impact on Sommerfeld's teaching and research. The Munich theorists were certainly aware of the momentousness of quantum mechanics because Sommerfeld remarked in a letter to Wentzel in January 1926: "I, too, believe that one has to convert completely to Heisenberg's new mechanics."[1] But the feat of his former prodigy student aroused little excitement. It took over half a year after the

---

[1] Sommerfeld to Wentzel, 13 January 1926, DM, NL 89, 004. "Auch ich glaube, dass man restlos zu Heisenbergs neuer Mechanik übergehen muss."

© The Author(s), under exclusive license to Springer Nature Switzerland AG 2020
M. Eckert, *Establishing Quantum Physics in Munich*,
SpringerBriefs in History of Science and Technology,
https://doi.org/10.1007/978-3-030-62034-9_7

publication of Heisenberg's paper "On the quantum theoretical reinterpretation of kinematic and mechanical relations" for quantum mechanics to become a subject of a Munich colloquium. Surprisingly, the presentation on February 19, 1926, was not given by Sommerfeld nor one of his assistants. Rather, Sommerfeld's colleague Constantin Caratheodory in the mathematics department spoke on the subject.[2]

Unlike this lukewarm response to "Heisenberg's new mechanics," Sommerfeld's reaction to Erwin Schrödinger's wave mechanics[3] was immediate and most enthusiastic. "This is terribly interesting," Sommerfeld responded when Schrödinger sent him his manuscript on "Quantization as an Eigenvalue Problem" prior to its publication:

> I was just busy preparing a concept for lectures in London (this March) that sing to an older tune. Then, like thunder, your manuscript arrived. My impression is this: your method replaces the new quantum mechanics of Heisenberg, Born, Dirac (R. Soc. Proc. 1925) and namely a simplified, analytic solution to their algebraic problem. Your results agree entirely with theirs.[4]

The same day Sommerfeld informed Pauli that Schrödinger apparently arrived at the "very same results as Heisenberg and you but in a quite different, totally crazy way. Instead of matrix algebra, he uses boundary value problems" (Sommerfeld 1926, 3).[5] Pauli shared Sommerfeld's enthusiasm. "I believe that this work belongs to the most important things that have been written recently," he wrote to Pascual Jordan in April 1926, shortly after Schrödinger's paper had first appeared. "Read it carefully and with devotion."[6] Sommerfeld avidly promoted Schrödinger's findings at his lectures in London in March 1926. "Schrödinger arrives at the same results as those obtained by the mechanics inaugurated by Heisenberg, but by a road that is presumably far simpler and more convenient [...] [Schrödinger's] treatment is expressed in the language of

---

[2]Physikalisches Mittwoch-Colloquium, DM, 1997–5115. Heisenberg visited Munich during the Easter holidays in April 1926 and presented a colloquium on "Magnetic electrons, complex structure and anomalous Zeeman effect" rather than on quantum mechanics. Obviously, he adjusted to the prevailing interest in Sommerfeld's institute; a week later, Wentzel talked about "The multiply periodic systems in the new quantum mechanics."

[3]Schrödinger to Sommerfeld, 29 January 1926, DM, NL 89, 013. Also in ASWB II. "[...] Ich bin natürlich auf niemandes Urteil so gespannt wie auf das Ihre, ob Sie die sehr hochgespannten Hoffnungen teilen, die ich an die Ableitung der Quantenvorschriften aus einem Hamiltonschen Prinzip knüpfe."

[4]Sommerfeld to Schrödinger, 3 February 1926, DM, NL 89, 004. Also in ASWB II. "Das ist ja furchtbar interessant [...] Ich war gerade dabei, für Vorträge in London (diesen März) ein Konzept zu machen, das aus der früheren Tonart blies. Da traf, wie ein Donnerschlag, Ihr Manuskript ein. Mein Eindruck ist dieser: Ihre Methode ist ein Ersatz der neuen Quantenmechanik von Heisenberg, Born, Dirac (R. Soc. Proc. 1925) und zwar ein vereinfachter, sozusagen eine analytische Resolvente des dort gestellten algebraischen Problems. Denn Ihre Resultate stimmen ganz mit jenen überein."

[5]Sommerfeld to Pauli, 3 February 1926, DM, NL 89, 003. Also in ASWB II. "Schr. scheint ganz dieselben Resultate zu finden, wie Heisenberg und Sie aber auf einem ganz anderen, total verrückten, Wege, keine Matrixalgebra, sondern Randwertaufgaben."

[6]Pauli to Jordan, 12 April 1926. WPWB I. "Ich glaube, daß diese Arbeit mit zu dem Bedeutendsten zählt, was in letzter Zeit geschrieben wurde. Lesen Sie sie sorgfältig und mit Andacht."

the theory of vibrations." He illustrated the breakthrough with Landé's g-formula, which had posed considerable problems for the old quantum theory:

> This new mechanics also accounts for the circumstance that wherever j^2 would be expected in the old mechanics, $j(j + 1)$ actually occurs. I should further like to remark as a curious fact that in the papers by Schrödinger mentioned above the quantum product $n(n + 1)$ enters in exactly the same way as the expression $n(n + 1)$ in the differential equation of spherical harmonics. (Sommerfeld 1926, 12)

Schrödinger thanked Sommerfeld for making his theory known in England.[7] In contrast to matrix mechanics, wave mechanics also became a preferred subject of colloquium presentations in Munich during the summer semester of 1926.[8] "We keenly study Schrödinger's new Quantum theory here and hold it in very high esteem," Sommerfeld reported to England in June 1926.[9] In July, Sommerfeld invited Schrödinger to Munich to present his theory at the colloquium.[10] Heisenberg used the occasion to criticize wave mechanics, and the event turned into a showdown between the two theorists. "We had Schrödinger here, together with Heisenberg," Sommerfeld reported to Pauli. "My general impression is that wave mechanics is an admirable micromechanics, but that it cannot remotely solve fundamental quantum riddles."[11] Heisenberg's criticism was sobering. After the Munich showdown, there were two hostile quantum camps: the Göttingen matrix mechanics led by Heisenberg, Born, and Jordan on one side; and wave mechanics represented by Schrödinger on the other. Heisenberg found Schrödinger's theory "strange" and considered the reading of a recent paper by Pauli, whom he regarded as an ally, as a "true comeback after Schrödinger's Munich presentations."[12]

In Munich, however, the incident caused only passing irritation. "We believe in Heisenberg, but we calculate following Schrödinger's approach," Hans Bethe recalled of the attitude in Sommerfeld's institute in the summer of 1926 (preface Eckert and

---

[7]Schrödinger to Sommerfeld, 28 April 1926, DM, HS 1977–28/A, 314. Also in ASWB II. "Es ist furchtbar lieb von Ihnen, dass Sie mir schon in England Propaganda gemacht haben."

[8]Physikalisches Mittwoch-Colloquium, DM, 1997–5115. On 4 June Carathéodory and Wentzel lectured "Über Schrödingers Wellenmechanik" and "Über den Zusammenhang zwischen Schrödingers und Heisenbergs Quantenmechanik", on 2 July Wentzel and Albrecht Unsöld talked about "Wellenmechanik und Quantenbedingungen" and "Das Heliumproblem in der Wellenmechanik", respectively.

[9]Sommerfeld to Richardson, 12 June 1926. Richardson Papers, Harry Ransom Center, University of Texas at Austin.

[10]Physikalisches Mittwoch-Colloquium, DM, 1997–5115. Schrödinger reported on 23 July 1926 on "Grundgedanken der Wellenmechanik" and on the next day on "Neue Resultate der Wellenmechanik".

[11]Sommerfeld to Pauli, 26 July 1926. WPWB I. "Wir haben Schrödinger hier gehabt, zugleich mit Heisenberg. Mein allgemeiner Eindruck ist der, dass die "Wellenmechanik" zwar eine bewundernswürdige Mikromechanik ist, dass aber die fundamentalen Quantenrätsel dadurch nicht im Entferntesten gelöst werden."

[12]Heisenberg to Pauli, 28 July 1926. WPWB I. "[…] Lektüre war mir eine wahre Erholung nach Schrödingers Vorträgen hier in München."

Pricha 1984).[13] Bethe had arrived in Munich in the spring of 1926 as a student in his fifth semester at just the right time to learn wave mechanics in Sommerfeld's seminar. "So he made Schrödinger's set of five papers the theme of the seminar for the summer semester (May through July) 1926. Every participant had to report on some section (maybe 5 to 10 percent) of Schrödinger's papers" (Bethe 2000).

During this time, wave mechanics was assigned as a theme of a doctoral thesis for the first time in the Munich nursery. "At first, Sommerfeld proposed the wave mechanical treatment of the problem of two centers," Albrecht Unsöld recalled many years later on how he became the first student to specialize in wave mechanics. "I soon realized that this was not possible and began to work on things spectroscopic, which were more amenable. When I presented my work to Sommerfeld, he became really angry at first. But when he saw that I had found some new methods and theorems in the realm of spherical harmonics, he marked my work with the best grade."[14] In January 1926, Sommerfeld asked Unsöld to scrutinize the available data on the fine structure of hydrogen and compare it with the results of the old quantum theory; in June 1926, he repeated this study within the framework of wave mechanics (Unsöld and Sommerfeld 1926a, b).[15] "He is highly gifted," Sommerfeld praised of Unsöld in the report to the faculty about Unsöld's doctoral thesis. Unsöld combined mathematical virtuosity with physical insight by providing a proof that the action of electronic shells on exterior points is spherically symmetrical. To this end, Unsöld made use of the addition theorem of spherical harmonics."[16] With reference to Unsöld's dissertation, Sommerfeld subsequently designed "quantum mechanical atomic models" for the German Museum (Eckert 2009).

The advent of quantum mechanics coincided with the arrival of foreign students, mostly from the United States, who came to Europe to study. Foreign students had often chosen Sommerfeld's institute for their studies,[17] but the rise of international

---

[13]"Es war charakteristisch für Sommerfeld, daß er sein ganzes Theoretisches Seminar im Sommersemester 1926 den Schrödingerschen Arbeiten widmete. Es war ihm sofort klar, daß diese in Zukunft die Grundlage für die Rechnungen in der Quantentheorie sein würden. 'Wir glauben an Heisenberg, aber wir rechnen nach Schrödinger', sagte er 1926."

[14]Unsöld to Eckert, 1 September 1982. Private Papers. "Sommerfeld hatte mir erst die wellenmechanische Behandlung des Zweizentrenproblems vorgeschlagen. Ich sah bald, daß das nicht ging und fing an, allerlei besser traktable spektroskopische Dinge zu behandeln. Als ich dann die Arbeit Sommerfeld vorlegte, wurde er erst richtig böse. Als er dann aber sah, daß ich im Bereich der Kugelfunktionen einige neue Methoden bzw. Theoreme gefunden hatte, benotete er die Arbeit mit Summa cum laude."

[15]"Gewisse frühere Angaben über die Intensität der Feinstrukturkomponenten werden zurückgezogen und durch die nach der Schrödingerschen Wellenmechanik berechneten Werte ersetzt."

[16]Sommerfeld, report to the faculty, 11 December 1926, UAM, OC–I–52p. "Während seiner Studienzeit hat er 2 kleinere Abhandlungen aus dem Gebiete der Spektroskopie veröffentlicht, die ich der Doktorarbeit beilege. Er ist hochbegabt und berechtigt zu grossen Hoffnungen […] behandelt Verfasser im 2. Teil zunächst das Additionstheorem der Kugelfunktion und zeigt mit seiner Hilfe, dass die Wirkung der Elektronenschalen auf äussere Punkte einfache Kugelsymmetrie besitzt."

[17]For example, Demetrios Hondros from Greece (1909), Frederick W. Grover and Walter F. Colby from the United States (1909, 1912–1913), Brillouin from France (1912–1913), J. W. Fisher from England (1922), to name only those who stayed for at least one semester.

fellowship programs in the mid-1920s changed the application process and the destination was no longer based on the student's personal preference. Professors at the student's home university, professors at host institutes, and the representatives of the foundations (such as the International Education Board of the Rockefellar Foundation) negotiated each student's study abroad location. (Coben 1971; Sopka 1980; Seidel 1981; Assmus 1993). The choice of a host institute, therefore, required an agreement among at least three parties and reflected, in particular, the international renown of the host institute. As the author of *Atombau und Spektrallinien* and with the reputation of a charismatic teacher, Sommerfeld was admired by both students and foundation officers. After his sojourn in the United States in 1922–1923, Sommerfeld's institute became an even more attractive destination. In 1926, the Munich nursery hosted several American fellows. Among them were two brothers, Victor and Ernst Guillemin, who were enticed to study in Munich after attending Sommerfeld's lectures in Madison. "I was fortunate in making his [Sommerfeld's] personal acquaintance during that year [1923] and I became aware of his outstanding capabilities as both a great scientist and a great teacher," Victor Guillemin recalled of his motivation to go to Munich. Guillemin went on to write one of the very first quantum mechanical doctoral dissertations supervised by Sommerfeld. "Consequently quantum mechanics has been to me, not something I read about; I was there' when it was born."[18]

Another American research fellow who arrived in Munich in the summer of 1926 was Linus Pauling. He had just completed his Ph.D. in physical chemistry at CalTech in Pasadena and asked both Bohr and Sommerfeld whether they would accept him if he received a fellowship from the Guggenheim Foundation to conduct research in Europe. "Sommerfeld answered the letter and Bohr didn't," he later recalled of his reason to go to Munich. "The exciting thing to me were the lectures Sommerfeld was giving on Schrödinger quantum mechanics and of course the seminars were devoted to it." Sommerfeld involved Pauling immediately in advanced research. At first, he suggested the spinning electron as a subject for wave-mechanical treatment. "Well, it didn't appeal to me very much; at any rate I didn't get anything out of it."[19] Pauling instead proposed another problem more to his liking and yielded a successful outcome. In December 1926, Sommerfeld used his privilege as a corresponding member of the Royal Society in London to submit Pauling's paper for publication in the *Proceedings of the Royal Society*.[20]

Sommerfeld's enthusiasm for Schrödinger's wave mechanics was further sparked by personal visits and an exchange of advanced students. At Schrödinger's request, Sommerfeld arranged for Walter Heitler and Fritz London to spend research sojourns

---

[18]Quoted in (Sopka 1980, 2.41). See also (Guillemin 1926). His sojourn was funded by the Sheldon Fellowship. Viktor's brother, Ernst Guillemin, chose a theme from electrical engineering as the subject of his dissertation. See Sommerfeld's report to the faculty, 7 July 1926, UAM, OC–I–52p.

[19]Interview with Pauling by Heilbron, 27 March 1964, https://www.aip.org/history-programs/niels-bohr-library/oral-histories/3448, accessed 17 April 2020.

[20]Sommerfeld to Rutherford, 13 December 1926, DM, NI, 89, 003. See also (Pauling 1927).

as Rockefeller fellows with Schrödinger in Zurich.[21] London had studied with Sommerfeld before Ewald lured him away to Stuttgart as his assistant; Heitler completed his doctoral work under Herzfeld's supervision in the summer of 1926. "Well, then, of course, in Zurich I met London, which was to be a decisive turning point in my career," Heitler recalled of how he and London came to perform the wave-mechanical calculation of the covalent bond that made their names famous in Schrödinger's institute in Zurich.[22] With these early successes, wave or quantum mechanics (term usage depended more on personal idiosyncrasies than on the methods used) was turned into a tool for solving problems with atoms, molecules and solids—in that order. Even Heisenberg and Pauli, who had created their own nurseries of theoretical physics at the university in Leipzig in 1927 and the ETH Zurich in 1928, respectively, paid tribute to quantum-mechanical problem solving despite their craving for clarity in fundamental matters (James and Joas 2015). In the same vein, Heisenberg conceived a quantum-mechanical theory of ferromagnetism that involved a "resonance" interaction between spinning electrons.[23] Heisenberg did not elaborate upon his "resonance" idea until 1928 when Heitler and London had discerned quantum-mechanical exchange forces, such as the cause for chemical bonding (Heisenberg 1928). In 1927, Sommerfeld further extended the range of applications by amalgamating the classical Drude-Lorentz electron gas concept with the new Fermi-Dirac statistics into a semiclassical electron theory of metals. This theory paved the way for the modern quantum theory of solids. For a number of years, Sommerfeld's advanced students and American research fellows began their careers with a research problem in this specialty (Eckert 1987; Hoddeson et al. 1992; Joas and Eckert 2017).

By the late 1920s, about a dozen of Sommerfeld's disciples held chairs for theoretical physics elsewhere in Germany and abroad.[24] Some of these formed a sub-network centered in Munich with a focus on quantum mechanical applications. Advanced students who were trained in the involved quantum mechanical techniques were lured away from one center to another: "So you are trying to steal assistants," Sommerfeld remarked after Heisenberg had expressed an interest in Unsöld in a letter. In this case, Sommerfeld had already arranged a research trip for Unsöld as a Rockefeller

---

[21] Schrödinger to Sommerfeld, 28 April 1926, DM, HS 1977–28/A, 314. "Würden Sie bereit sein, mir Herrn London nächstes Jahr mit Rockefeller nach Zürich zu geben? Würde er wohl wollen? Und würden Sie es gut finden? Oder wüßten Sie jemand anderen?"; Schrödinger to Sommerfeld, 11 May 1926, DM, HS 1977–28/A, 314. "Oh das wäre furchtbar lieb, wenn Sie uns die Freude machten, nach Zürich zu kommen. Bitte tun Sie es wirklich! Wenn Sie nichts weiter schreiben, rechnen wir damit, Ihr Zimmer wird am 28. gerichtet sein […] Ich bin natürlich sehr froh, wenn irgendjemand von Ihnen zu mir kommt, auch Heitler."

[22] Interview with Heitler by Heilbron, 18 March 1963, https://www.aip.org/history-programs/niels-bohr-library/oral-histories/4662-1, accessed 17 April 2020. See also (Heitler and London 1927).

[23] Heisenberg to Pauli, 4 November 1926. Quoted in (Hoddeson et al. 1992, 129–135).

[24] Epstein in Pasadena (California); Ewald in Stuttgart; H. G. Grimm in Würzburg; Heisenberg in Leipzig; Herzfeld in Baltimore (Maryland); Kratzer in Münster; Laporte in Ann Arbor (Michigan, United States); Max von Laue in Berlin, Lenz in Hamburg; Heinrich Ott in Würzburg; Pauli in Zurich (ETH); Pauling in Pasadena; Rudolf Seeliger in Greifswald; Wentzel in Zurich.

fellow at the Mount Wilson Observatory to foster a connection with astrophysics.[25] Pauli, too, was eager to have "a reasonable quantum man" as an assistant.[26] Pauli wrote to Sommerfeld in May 1929:

> Now we are not very far away from one another and I hope for closer contact between Zurich and Munich physics [...] I now have a fairly large operation here in Zurich. Mr. Bloch is busy elaborating a theory of supraconductivity [...] Mr. Peierls is dealing with the theory of heat conductivity in solids."[27]

Peierls was shared among the three centers. When Pauli requested Peierls as his assistant in 1929, Heisenberg agreed under the provision that Pauli, too, send him good physicists in Leipzig from time to time. "It would be very nice if we could establish a kind of exchange of physicists between Zurich and Leipzig."[28]

References

Assmus A (1993) The Creation of Postdoctoral Fellowships and the Siting of American Scientific Research Students. Minerva 31:151–183
Bethe HA (2000) Sommerfeld's Seminar. Phys Perspect 2:3–5
Coben S (1971) The Scientific Establishment and the Transmission of Quantum Mechanics to the United States, 1919–32. Am Hist Rev 76:442–466
Eckert M (1987) Propaganda in Science: Sommerfeld and the Spread of Electron Theory of Metals. Hist Stud Phys Sci 17(2):191–233
Eckert M (2009) Quantenmechanische Atommodelle zwischen musealer Didaktik und ideologischer Auseinandersetzung. In: Bigg C, Hennig J (eds) Atombilder: Ikonografie des Atoms in Wissenschaft und Öffentlichkeit des 20. Jahrhunderts. Wallstein Verlag, Göttingen, pp 83–91
Eckert M, Pricha W (1984) Boltzmann, Sommerfeld und die Berufungen auf die Lehrstühle für theoretische Physik in Wien und München, 1890–1917. Mitteilungen der Österreichischen Gesellschaft für Geschichte der Naturwissenschaften 4:101–119

---

[25]Sommerfeld to Heisenberg, 15 November 1927. DMA, NL 89, 002. "Ihr Brief atmet von Anfang bis zu Ende Ihr schlechtes Gewissen aus. Sie wollen also Assistenten stehlen? und natürlich gerade die besten! Ich habe ein besseres Gewissen wie Sie und habe Ihren Brief Unsöld gezeigt. Wenn er auch grosse Lust hätte, zu Ihnen zu kommen, so stimmt er mir doch vollkommen bei, dass er jetzt sozusagen mitten im Semester nicht fortgehen kann, (er ist nämlich auch hier für das Seminar sehr nützlich) und dass es für ihn selbst auf die kurze Zeit bis 1. April, wo er nach Amerika gehen soll, keinen Zweck hat seine Tätigkeit zu wechseln."

[26]Pauli to Ralph Kronig, 7 February 1928, WPWB I. "[...] einen vernünftigen Quantenmann [...]."

[27]Pauli to Sommerfeld, 16 May 1929, DM, HS 1977–28/A,254. Also in ASWB II. "Nun sind wir ja nicht sehr weit voneinander entfernt und ich hoffe in Zukunft auf einen engeren Kontakt zwischen der Züricher und der Münchener Physik [...] Ich habe jetzt einen ziemlich großen Betrieb hier in Zürich. Herr Bloch ist zur Zeit mit der Ausarbeitung einer Theorie der Supraleitung beschäftigt. Die Sache ist noch nicht fertig, scheint aber zu gehen. Herr Peierls treibt Theorie der Wärmeleitung in festen Körpern."

[28]Heisenberg to Pauli, 1 August 1929, WPWB I. "Du willst also im nächsten Semester den Peierls als Assistenten haben? Mir ist das natürlich im Prinzip völlig recht; ich finde aber, Du mußt mir als Ersatz auch gute Physiker nach L[eipzig] schicken; besonders möchte ich gern, daß der Bloch wieder für eine Zeit nach L[eipzig] kommt, ließe sich das nicht machen? Ich fände sehr schön, wenn wir so eine Art Physikeraustausch zwischen Zürich und L[eipzig] einrichten könnten [...]."

Guillemin V (1926) Zur Molekülstruktur des Methan. Annalen der Physik 81:173–204

Heisenberg W (1928) Zur Theorie des Ferromagnetismus. Zeitschrift für Physik 49:619–636

Heitler W, London F (1927) Wechselwirkung neutraler Atome und homöopolare Bindung nach der Quantenmechanik. Zeitschrift für Physik 44:455–472

Hoddeson L, Baym G, Eckert M (1992) The Development of the Quantum Mechanical Electron Theory of Metals, 1926–1933. In: Hoddeson L et al (eds) Out of the Crystal Maze. Chapters from the History of Solid-State Physics. Oxford University Press, New York, pp 88–181

James J, Joas C (2015) Subsequent and Subsidiary? Rethinking the Role of Applications in Establishing Quantum Mechanics. Hist Stud Nat Sci 45:641–702

Joas C, Eckert M (2017) Arnold Sommerfeld and Condensed Matter Physics. Annu Rev Condens Matter Phys 8:31–49

Pauling L (1927) Refraction, Diamagnetic Susceptibility, and Extension in Space. Proc R Soc A 114:181–211

Seidel RW (1981) Aspetti istituzionali della trasmissione della meccanica quantistica agli Stati Uniti. In: de Maria M et al (eds) Fisica & Società negli anni '20. LLUP-CLUED, Milan, pp 189–214

Sommerfeld A (1926) Three Lectures on Atomic Physics. Methuen, London

Sopka KR (1980) Quantum Physics in America. Arno Press, New York

Unsöld A, Sommerfeld A (1926a) Über das Spektrum des Wasserstoffes. Zeitschrift für Physik 36:259–275

Unsöld A, Sommerfeld A (1926b) Über das Spektrum des Wasserstoffes: Berichtigungen und Zusätze. Zeitschrift für Physik 38:236–241

Chapter 8
Conclusion

Abstract Arnold Sommerfeld's approach to quantum problems displays twists and turns that do not follow a common trend. Quantum theory appeared intermittently and in different guises as Sommerfeld developed his institute into a "nursery" of theoretical physics. Nevertheless, the choice of themes and methods was not arbitrary. There was a marked preference for solving problems amenable to the skilled use of mathematical tools. Agreement with empirical results was considered more important than adherence to principles. This attitude was perfectly suited to accommodating quantum theory as a research subject of his school when physics underwent a period of rapid change.

Keywords Physics of problems

In addition to the analysis of Sommerfeld's performance as a theorist (Seth 2010), the development of the Bohr-Sommerfeld theory prior to quantum mechanics (Kragh 2012), and the biographical study of Sommerfeld's life (Eckert 2013), this book addressed the institutional context of Sommerfeld's Munich quantum nursery. What were the emergent features of the Munich School in the course of the two decades after Sommerfeld was called there as Boltzmann's successor? As is obvious from Table A.1, Sommerfeld did not have to start from scratch when he founded his nursery. For the first few years, he adapted his pedagogical practice to the routine of Munich lectures as it had developed in the interregnum between Boltzmann's leave in 1894 and Sommerfeld's arrival in 1906. Sommerfeld's canonical six-semester lecture courses gradually took shape around 1910. In the beginning, the "seminar" was only an exercise that accompanied Sommerfeld's main lectures. Like Sommerfeld's advanced lectures, the seminar did not immediately acquire the character of a research-oriented performance addressed to advanced students—as recalled by the prodigal quantum students from the 1920s. If we take the lectures and seminars as an expression of Sommerfeld's pedagogical practice, quantum theory appeared intermittently and in different guises, adapting to the needs and opportunities of the time.

By the same token, Sommerfeld's approach to quantum problems displays twists and turns that do not follow a common trend. Sommerfeld's h-hypothesis at the time

M. Eckert, *Establishing Quantum Physics in Munich*,
SpringerBriefs in History of Science and Technology,
https://doi.org/10.1007/978-3-030-62034-9_8

of the First Solvay Conference in 1911, his extension of Bohr's model in 1916, the demise of models in the course of the study of complex spectra in the early 1920s, and the focus on wave mechanics after 1926. Each of these appear contradictory when we expect a common, unifying factor. Nevertheless, Sommerfeld's choice of themes and methods was not arbitrary. Suman Seth's distinction between "physics of problems" versus "physics of principles" exemplified by Sommerfeld and Planck, respectively, aptly displays different features in the way one teaches and performs as a theoretical physicist (Seth 2010). However, Sommerfeld's "physics of problems" should not be mistaken for a hodgepodge of incoherent problem solving. We may discern marked preferences, such as a choice of problems suited to further treatment using Sommerfeld's skilled use of partial differential equations. Whether a subject was considered an appropriate target, however, depended not only on the availability of suitable mathematical tools, but also on empirical results that would serve as a proving ground for the theory. Sommerfeld's h-hypothesis emerged from his ongoing interest in X-rays, for example; his awareness of new experimental discoveries is also illustrated by his response to the Paschen-Back and Stark effects in 1913. Sommerfeld's approach may have displayed a lack of the physics of principles— but the same craving for theories in agreement with empirical results and epistemic openness was perfectly suited to accommodate quantum theory as a research subject for many disciples in his nursery.

References

Eckert M (2013) Arnold Sommerfeld: Atomphysiker und Kulturbote 1868–1951. Eine Biografie. Göttingen: Wallstein. Abhandlungen und Berichte des Deutschen Museums, Neue Folge, Bd. 29. American Translation: Arnold Sommerfeld: Science, Life and Turbulent Times 1868–1951. Springer, New York
Kragh H (2012) Niels Bohr and the Quantum Atom: The Bohr Model of Atomic Structure 1913–1925. Oxford University Press, Oxford
Seth S (2010) Crafting the Quantum. Arnold Sommerfeld and the Practice of Theory, 1890–1926. The MIT Press, Cambridge, MA

Appendix

(See Tables A.1, A.2, A.3, A.4 and A.5).

Table A.1 Munich lectures on theoretical physics, 1904–1914. (SS = Summer Semester; WS = Winter Semester; G = Graetz; K = Korn; D = Donle; S = Sommerfeld; the number in brackets means hours per week.)

SS 1904	G: Theorie des Lichts (4)
	G: Einleitung in die theoretische Physik (4)
	K: Einführung in die analytische Mechanik (4)
	K: Über die Telegraphengleichung und die Theorie der Wechselströme (2)
	D: Doppelbrechung und damit zusammenhängende Erscheinungen (2)
WS 1904/05	G: Analytische Mechanik (5)
	G: Theorie der Elektronen (2)
	K: Die partiellen Differentialgleichungen der mathematischen Physik (4)
	D: Einführung in die elektromagnetische Theorie des Lichtes (2)
SS 1905	G: Analytische Mechanik II (4)
	G: Theorie der Wärme und kinetische Gastheorie (4)
	K: Einführung in die analytische Mechanik (4)
	K: Über die Telegraphengleichung und die Theorie der Wechselströme (2)
	D: Interferenz- und Beugungserscheinungen in der Optik (2)
WS 1905/06	G: Maxwellsche Theorie der Elektrizität und des Magnetismus (5)
	G: Theoretische Akustik (2)
	K: Variationsrechnung, mit Rücksicht auf die Anforderungen der theoretischen Physik (2)
	K: Kinetische Gastheorie (2)
	D: Physikalische Maße und Messmethoden (2)

(continued)

© The Author(s), under exclusive license to Springer Nature Switzerland AG 2020
M. Eckert, *Establishing Quantum Physics in Munich*,
SpringerBriefs in History of Science and Technology,
https://doi.org/10.1007/978-3-030-62034-9

Table A.1 (continued)

SS 1906	G: Einleitung in die theoretische Physik (4)
	G: Theorie des Lichts (4)
	K: Funktionentheorie nach Cauchy und Riemann; ihre Anwendungen in der theoretischen Physik (4)
	D: Einführung in die neuere Elektrizitätslehre (2)
WS 1906/07	G: Analytische Mechanik (5)
	K: Potentialtheorie und Kugelfunktionen (4)
	D: Einführung in die elektromagnetische Theorie des Lichtes (2)
	S: Maxwellsche Theorie und Elektronentheorie (4)
SS 1907	S: Theorie der Strahlung (3)
	S: Grundzüge der Thermodynamik (2)
	S: Übungen (2)
	G: Einleitung in die theoretische Physik (4)
	G: Analytische Mechanik II (2)
	D: Doppelbrechung und damit zusammenhängende Erscheinungen (2)
WS 1907/08	S: Kinetische Gastheorie (3)
	S: Ausgewählte Fragen der Thermodynamik (2)
	S: Seminar über thermodynamische und thermochemische Fragen (2)
	G: Theorie der Elektrizität und des Magnetismus (5)
	G: Über die Fortschritte der exakten Naturwissenschaften (1)
	D: Physikalische Maße und Messmethoden (2)
SS 1908	S: Wärmeleitung, Diffusion und Elektrizitätsleitung (3)
	S: Hydrodynamik für Anfänger (2)
	S: Seminar über hydrodynamische Fragen (2)
	G: Einleitung in die theoretische Physik (4)
	G: Elektromagnetische Lichttheorie (3)
	D: Einführung in die neuere Elektrizitätslehre (2)
WS 1908/09	S: Elektrodynamik, insbesondere Elektronentheorie (4)
	S: Über die Bedeutung der Kreiseltheorie für die allgemeine Mechanik und Physik (2)
	S: Seminar über elektrodynamische Fragen (2)
	G: Analytische Mechanik (4)
	D: Einführung in die elektromagnetische Theorie des Lichtes (2)
SS 1909	S: Optik (4)
	S: Seminar über optische Fragen (2)
	G: Einleitung in die theoretische Physik (3)
	D: Doppelbrechung und damit zusammenhängende Erscheinungen (2)

(continued)

Table A.1 (continued)

WS 1909/10	S: Vektoranalysis, als Einleitung in die mathematische Physik (3)
	S: Thermodynamik (3)
	S: Vektoranalysis, mit Übungen (3)
	S: Seminar, Vorträge der Mitglieder und Demonstrationen (2)
	G: Theorie der Elektrizität und des Magnetismus (3)
	D: Physikalische Maße und Messmethoden (2)
	von Laue: Interferenz und Beugung (1)
SS 1910	S: Partielle Differentialgleichungen der Physik für Anfänger (3)
	S: Kinetische Gastheorie und Verwandtes (3)
	S: Übungen zu Partielle Differentialgleichungen (mit von Laue) (1)
	S: Seminar: Vorträge der Mitglieder über die statistischen Methoden (2)
	G: Einleitungen die theoretische Physik (3)
	D: Einführung in die neuere Elektrizitätslehre (2)
	von Laue: Besprechung von Arbeiten aus dem Gebiete der Relativitätstheorie (Vorträge der Teilnehmer) (2)
WS 1910/11	S: Analytische Mechanik (4)
	S: Geometrische Optik (2)
	S: Seminar: Übungsaufgaben zur Mechanik (2)
	G: Einleitung in die theoretische Physik II (2)
	D: Einführung in die elektromagnetische Theorie des Lichtes (2)
	von Laue: Vektoranalysis (2)
	von Laue: Besprechung über die Anwendung der Thermodynamik auf chemische Fragen (Vorträge der Teilnehmer) (2)
SS 1911	S: Mechanik der Kontinua: Hydrodynamik, Akustik, Elastizität für Anfänger (4)
	S: Elektrodynamik und Mechanik vom Standpunkt des Relativitätsprinzips (3)
	S: Seminar: Übungsaufgaben zu Mechanik der Kontinua (2)
	S: Seminar: Vorträge der Mitglieder über Relativität (2)
	G: Einleitung in die theoretische Physik (3)
	D: Doppelbrechung und damit zusammenhängende Erscheinungen (2)
	von Laue: Thermodynamik (mit besonderer Berücksichtigung der chemischen Anwendungen) (2)
	Debye: Einführung in die theoretische Optik (mit Demonstrationen) (4)
	Debye: Elektronenoptik (Zeeman-Effekt, lichtelektrische Wirkungen usw.) (1)
WS 1911/12	S: Maxwellsche Theorie, Grundlagen und einfachere Teile derselben (4)
	S: Theorie der Röntgenstrahlen und Verwandtes, für Vorgeschrittene (2)
	S: Seminar: Übungsaufgaben zur Maxwellschen Theorie (2)
	G: Einleitung in die theoretische Physik, II (2)
	D: Physikalische Maße und Messmethoden (2)
	von Laue: Theoretische Optik (3)

(continued)

Table A.1   (continued)

SS 1912	S: Thermodynamik, einschließlich kinetischer Gastheorie (5)
	S: Elektronentheorie der Metalle und statistische Fragen, für Vorgeschrittene (2)
	S: Seminar: Übungsaufgaben zur Thermodynamik (2)
	D: Einführung in die theoretische Elektrizitätslehre (2)
	von Laue: Das Relativitätsprinzip und seine Folgerungen (3)
WS 1912/13	S: Theoretische Optik nebst Einführung in ihre elektromagnetischen Grundlagen (4)
	S: Ausgewählte Fragen aus der Quantentheorie, für Vorgeschrittenere (2)
	S: Seminar: Demonstrationen und Übungsaufgaben zur Optik (2)
	G: Einleitung in die theoretische Physik I (3)
	D: Einführung in die elektromagnetische Theorie des Lichtes (2)
SS 1913	S: Partielle Differentialgleichungen der Physik (4)
	S: Relativitätstheorie (2)
	S: Übungen (2)
	G: Einleitung in die theoretische Physik II (2)
	D: Doppelbrechung und damit zusammenhängende Erscheinungen (2)
WS 1913/14	S: Mechanik (4)
	S: Ausgewählte Fragen der Statistik für Vorgeschrittenere (2)
	S: Übungsaufgaben (2)
	G: Einleitung in die theoretische Physik I (3)
	D: Physikalische Maße und Meßmethoden (2)
SS 1914	S: Mechanik der Kontinua (Hydrodynamik, Akustik, Elastizität) (4)
	S: Besprechungen zur Quantentheorie (2)
	S: Übungen zur Hydrodynamik (2)
	G: Einleitung in die theoretische Physik II (3)
	D: Einführung in die theoretische Elektrizitätslehre (2)
WS 1914/15	S: Elektrodynamik (Maxwellsche und Elekronentheorie) (4)
	S: Zeeman-Effekt und Spektrallinien (1)
	S: Seminar: Übungen zur Maxwellschen Theorie (2)
	G: Einleitung in die theoretische Physik I (3)
	D: Einführung in die elektromagnetische Theorie des Lichtes (2)
	Lenz: Einführung in die Potentialtheorie (2)
	Lenz: Wärmestrahlung (2)

Table A.2 Sommerfeld's doctoral students (1908–1928)

1908	Frederick W. Grover	Die Methode der Induktionswaage
	Peter Debye	Der Lichtdruck auf Kugeln von beliebigem Material
1909	Demetrios Hondros	Über elektromagnetische Drahtwellen
	Ludwig Hopf	Turbulenzerscheinungen an einem Fluß. Schiffswellen
1910	Rudolf Seeliger	Beitrag zur Theorie der Elektrizitätsleitung in dichten Gasen
1911	Wilhelm Lenz	Über das elektromagnetische Wechselfeld der Spulen und deren Wechselstromwiderstand, Selbstinduktion und Kapazität
	Wilhelm Hüter	Kapazitätsmessungen an Spulen
	Herman W. March	Über die Ausbreitung der Wellen der drahtlosen Telegraphie auf der Erdkugel
	Harald von Hörschelmann	Über die Wirkungsweise des geknickten Marconischen Senders
1912	Paul P. Ewald	Dispersion und Doppelbrechung von Elektronengittern (Kristallen)
1914	Paul Epstein	Über die Beugung an einem ebenen Schirm unter Berücksichtigung des Materialeinflusses
	Alfred Landé	Zur Methode der Eigenschwingungen in der Quantentheorie
	Walter Dehlinger	Über die Spezifische Wärme zweiatomiger Kristalle
1917	Karl Glitscher	Spektroskopischer Vergleich zwischen den Theorien des starren und des deformierbaren Elektrons
1918	Franz Pauer	Magnetische Drehung der Polarisationsebene des Lichtes in einem Gase Bohrscher Moleküle
1919	Herbert Lang	Zur Tensorgeometrie in der allgemeinen Relativitätstheorie
	Erwin Fues	Vergleich zwischen den Funkenspektren der Erdalkalien und den Bogenspektren der Alkalien
1920	Walter Heine	Untersuchung aus dem Gebiet der Kristallstruktur. I. Die quadratischen Formen der Bravais'schen Gitter. II. Strukturtheorie der accidentellen Doppelbrechung bei Steinsalz
	Josef Krönert	Gesetzmässigkeiten beim anomalen Zeemaneffekt
	Adolf Kratzer	Zur Quantentheorie der Rotations-Spektren
1921	Walter Schaetz	Über die Druckempfindlichkeit einer kreisförmigen elastischen Membran
	Gregor Wentzel	Zur Systematik der Röntgenspektren
	Josef Weinacht	Zur Kritik der Theorie des erweiterten Rydberg-Ritzschen Serienterms
	Wolfgang Pauli	Über das Modell des Wasserstoff-Molekülions
1922	Johannes Fischer	Über die Beugungserscheinungen bei sphärischer Aberration
1923	Werner Heisenberg	Über Stabilität und Turbulenz von Flüssigkeitsströmen
1924	Heinrich Ott	Röntgenographische Untersuchungen
	Otto Laporte	Die Struktur des Eisenspektrums

(continued)

Table A.2 (continued)

1925	Karl Bechert	Die Struktur des Nickelspektrums
1926	Helmut Hönl	Zum Intensitätsproblem der Spektrallinien
	Walter Heitler	Zwei Beiträge zur Theorie konzentrierter Lösungen
	Ernst Guillemin	Zur Theorie der Frequenzvervielfachung durch Eisenkernkoppelung
	Viktor Guillemin	Zur Molekülstruktur des Methan
1927	Albrecht Unsöld Hellmut Seyfarth	Beiträge zur Quantenmechanik der Atome Röntgenstrukturananlyse
1928	Hans Bethe	Theorie der Beugung von Elektronen an Kristallen
	Hermann Brück	Über die wellenmechanische Berechnung von Gitterkräften und die Bestimmung von Jonengrößen, Kompressibilitäten und Gitterenergien bei einfachen Salzen
	Johann Gratsiatos	Über das Verhalten der radiotelegraphischen Wellen in der Umgebung des Gegenpunktes der Antenne und über die Analogie zu den Poissonschen Beugungserscheinungen
	August Kupper	Zur Frage der Intensität von Spektrallinien

Table A.3 Sommerfeld's assistants and *Privatdozenten* (1906–1928). (The year in the first row indicates when they began as an assistant to Sommerfeld; title and year of habilitation are given if it was supervised by Sommerfeld.)

1906	Peter Debye	Zur Theorie der Elektronen in Metallen (1910)
1909	Max von Laue	
1911	Walter Friedrich	
1911	Wilhelm Lenz	Berechnung der Eigenschwingungen einlagiger Spulen (1914)
1913	Paul P. Ewald	Zur Kristalloptik der Röntgenstrahlen (1917)
1914	Karl Glitscher	
1915	Adalbert Rubinowicz	
1919	Adolf Kratzer	Zur Theorie der Bandenspektren (1921)
1920	Karl Herzfeld	
1921	Gregor Wentzel	Zur Theorie der Streuung von β-Strahlen (1922)
1922	Heinrich Ott	Die Kristallstruktur des Graphits (1928)
1926	Karl Bechert	Die Intensitäten von Dublettlinien nach der Diracschen Theorie (1930)

Table A.4 Lectures by Sommerfeld and his assistants/*Privatdozenten* (1914–1928). (S = Sommerfeld; the lecture in the first line of each semester is the regular course lecture with four hours per week; the other lectures were usually two hours per week; L = Lenz; E = Ewald; H = Herzfeld; K = Kratzer; W = Wentzel; O = Ott.)

WS 1914/15	S: Elektrodynamik (Maxwellsche und Elektronentheorie)
	S: Zeeman-Effekt und Spektrallinien
	L: Einführung in die Potentialtheorie
	L: Wärmestrahlung
SS 1915	S: Optik
	S: Relativitätstheorie
WS 1915/16	S: Thermodynamik einschließlich kinetischer Gastheorie
	S: Probleme der Atomistik
SS 1916	S: Partielle Differentialgleichungen der mathematischen Physik
	S: Grundlagen der Elektrodynamik
WS 1916/17	S: Mechanik
	S: Quantentheorie (für Vorgeschrittenere)
	S: Neuere experimentelle und theoretische Fortschritte in der Atomistik und Elektronik (populär, ohne mathematische Entwickelungen)
SS 1917	S: Mechanik der Continua (Hydrodynamik, Akustik, Elastizität)
	S: Relativitätstheorie
WS 1917/18	S: Elektrodynamik (Maxwellsche Theorie und Elektronentheorie)
	S: Probleme der drahtlosen Telegraphie
	S: Röntgenstrahlen und Kristallstruktur, für Hörer aller Fakultäten, ohne mathematische Ableitung
SS 1918	S: Optik
	S: Atomistik (für Hörer aller Fakultäten)
WS 1918/19	S: Thermodynamik einschl. kinetischer Gastheorie
	S: Wiederholungskurs in Mechanik, einschließlich Hydrodynamik und Elastizität (gemeinsam mit Dr. Ewald)
	S: Quantentheorie
	S: Atombau und Spektrallinien
	L: Quantentheorie
	E: Vektoranalysis, als Einführung in die theoretische Physik
	L: Wiederholungskurs in Elektrodynamik
SS 1919	S: Partielle Differentialgleichungen der Physik
	S: Repetitorien für Kriegsteilnehmer, zusammen mit den Priv. Doz. Dr. Lenz und Dr. Ewald
	S: Relativitätstheorie
	E: Vektorrechnung als Grundlage für die theoretische Physik (mit Übungen)
	L: Mathematische Einleitung in die physikalische Chemie
	L: Quantentheorie

(continued)

Table A.4 (continued)

WS 1919/20	S: Optik
	S: Atombau und Spektrallinien (für Hörer aller Fakultäten)
	E: Dynamik der Kristallgitter
	L: Kinetische Gastheorie mit besonderer Berücksichtigung der Brownschen Bewegung
SS 1920	S: Elektrodynamik
	S: Theorie der Röntgenstrahlen (für Vorgeschrittene)
	S: Relativitätstheorie (für Hörer aller Fakultäten)
	E: Vektorrechnung (zur Einführung in die theoretische Physik)
	E: Ausgewählte Probleme der Elektronenoptik
	L: Relativitätstheorie
	H: Einführung in die statistische Mechanik
	H: Nernstsches Wärmetheorem
WS 1920/21	S: Mechanik
	S: Theorie d. Röntgenstrahlen (für Vorgeschrittene)
	S: Theorie d. Spektrallinien auf Grund des Bohrschcn Atommodelles (für Hörer aller Fakultäten)
	L: Theorie der Wärmestrahlung
	E: Thermodynamische Potentiale
	H: Elektronentheorie der Metalle
	H: Einführung in die statistische Mechanik
	H: Chemische Anwendungen der kinetischen Theorie und Quantentheorie
SS 1921	S: Hydrodynamik, Elastizität etc.
	S: Magneto- u. Elektro-Optik
	E: Kinetische Gastheorie
	H: Vektoranalysis
	H: Mathematische Einleitung in die physikalische Chemie
WS 1921/22	S: Maxwellsche Theorie u. Elektronentheorie
	S: Randwertaufgaben aus der Maxwellschen Theorie
	H: Quantenmechanik der Atommodelle
	H: Einführung in die kinetische Gastheorie
SS 1922	S: Optik
	S: Elektrooptik u. Magnetooptik
	H: Mathematische Einführung in die physikalische Chemie
	H: Magnetismus
	H: Nernstsches Wärmetheorem
	K: Vektoranalysis mit Anwendungen
	K: Theorie der Wärmestrahlung
WS 1922/23	H: Thermodynamik und kinetische Gastheorie (in Vertretung für Sommerfeld)

(continued)

Table A.4 (continued)

SS 1923	S: Partielle Differentialgleichungen
	S: Spektroskopische Probleme
	H: Zustandsgleichung
	W: Einführung in die theoretische Elektrizitätslehre
	W: Ausgewählte Kapitel der Kristallphysik
WS 1923/24	S: Mechanik
	S: Allgemeine Relativitätstheorie für Vorgeschrittene
	H: Theorie der Dispersion
	H: Aufgaben aus der theoretischen Physik, gemeinsam mit Wentzel
	W: Vektoranalysis (Einführung in die theoretische Physik)
	W: Durchgang von Kathodenstrahlen durch Materie (ausgewählte theoretische Probleme)
SS 1924	S: Mechanik der Continua (Hydrodynamik. Elastizität u.s.w.)
	S: Ausgewählte Fragen der Quantentheorie
	H: Höhere Mechanik
	H: Die Größe der Moleküle und Atome
	W: Theorie der Wärmestrahlung
	W: Aufgaben aus der theoretischen Physik
WS 1924/25	S: Maxwellsche Theorie der Elektrodynamik
	S: Probleme der Spektroskopie
	W: Vektoranalysis (Einführung in die theoretische Physik)
	W: Schwankungserscheinungen (Brownsche Bewegung)
SS 1925	S: Optik
	S: Dynamische Probleme der Atomphysik
	H: Oberflächenerscheinungen (Kapillarität und Adsorption vom atomistischen Standpunkt)
	W: Elektrodynamik bewegter Körper und Relativitätstheorie
WS 1925/26	S: Thermodynamik u. kinetische Gastheorie
	S: Ausgewählte Fragen der Quantentheorie
	H: Theorie der Kristallgitter
	W: Vektoranalysis (Einführung in die theoretische Physik)
	W: Theorie der Bandenspektren
	W: Zerstreuung von Licht und Röntgenstrahlen
SS 1926	S: Partielle Differentialgleichungen der Physik
	S: Über einige Grundfragen der Physik (für Hörer aller Fakultäten)
	S: Theorie des Magnetismus
	W: Theorie der Bandenspektren
	W: Zerstreuung von Licht und Röntgenstrahlen
WS 1926/27	S: Mechanik
	S: Ausgewählte Fragen der Quantentheorie
	W: Vektoranalysis (Einführung in die theoretische Physik)
	W: Optik der Kristalle

(continued)

Table A.4 (continued)

SS 1927	S: Mechanik der Continua (Hydrodynamik, Elastizität, Capillarität)
	S: Struktur der Materie
	O: Untersuchungsmethoden, Eigenschaften u. Theorie der Kristallgitter
WS 1927/28	S: Maxwellsche Theorie der Elektrodynamik
	S: Über Quantenmechanik (für Vorgeschrittene)
	O: Vektoranalysis u. Einführung in die theoretische Physik
	O: Kinetische Gastheorie
SS 1928	S: Optik
	S: Ausgewählte Fragen der Wellenmechanik (für Vorgeschrittene)
	O: Zerstreuung von Licht und Röntgenstrahlen

Table A.5 Research Fellows at Sommerfeld's institute (1924–1928). (The year in the first row is the year of arrival; each fellowship's duration varied between a month and more than a year.)

1924	Miguel Catalan	Die Struktur des Kobaltspektrums (mit Bechert, ZfP 1925)
		Über das Bogenspektrum des Palladiums (mit Bechert, ZfP 1926)
		Über einige allgemeinere Regelmäßigkeiten der optischen Spektren (mit Bechert, ZfP 1926)
1925	Viktor Guillemin	Zur Molekülstruktur des Methan (AdP 1926)
	Ernst Guillemin	Zur Theorie der Frequenzvervielfachung durch Eisenkernkoppelung (Archiv für Elektrotechnik 1926)
1926	Linus Pauling	The Theoretical Prediction of the Physical Properties of Many-Electron Atoms and Ions (Proc. Roy. Soc. 1927)
		Die Abschirmungskonstanten der relativistischen oder magnetischen Röntgenstrahlendubletts (ZfP 1927)
1927	Carl Eckart	Über die Elektronentheorie der Metalle auf Grund der Fermischen Statistik, insbesondere über den Volta-Effekt (ZfP 1928)
	William V. Houston	Elektrische Leitfähigkeit auf Grund der Wellenmechanik (ZfP 1928)
	Edwin C. Kemble	Die Elektronenemission kalter Metalle (ZfP 1928)
	Isidor I. Rabi	General Principles of quantum mechanics (Rev. Mod. Phys. 1929)
	Edward U. Condon W. W. Sleator	Das freie Elektron im homogenen Magnetfeld nach der Diracschen Theorie (ZfP 1928)
1928	Allan C. G. Mitchell	Entropie des Elektronengases auf Grund der Fermischen Statistik (ZfP 1928)

Printed in the United States
By Bookmasters